BODY WORLDS

ANIMAL INSIDE OUT

TABLE OF CONTENTS

Gunther von Hagens

Anatomical Safari

Humans are attracted to new and unusual stimuli from their environment. They make them curious and, in some cases, such stimuli even inspire wonder. "Wonder is the desire for knowledge", as philosopher Thomas Aquinas wrote. Wonder may result in both admiration and astonishment. Admiration makes humans look at nature in fascination; astonishment makes them ask how nature accomplishes all it does. The amazing variety in the builds of animal bodies causes both admiration and astonishment. Revealing the beauty of animal bodies is the task of veterinary anatomy that studies the form and structure of these bodies.

Thanks to the revolutionary technology of Plastination, it has become possible to dissect and permanently preserve even the largest animals in an aesthetically pleasing manner. ANIMAL INSIDE OUT, a Body Worlds Production is a journey through the substrata of animal life. It grants its visitors a glance underneath skin, fur and feathers in order to convey for the first time detailed impressions of the bone structure, muscles, nerves and organs of animals. On this anatomical safari, visitors learn in great clarity about the similarities that exist in all vertebrates and, at the same time, about the great anatomical variety that has evolved in the vertebrate group as well. Many questions are answered: how many muscles are there in an elephant's trunk? How do sharks breathe? How much does the heart of a giraffe weigh? ANIMAL INSIDE OUT is designed to enrich and inform people of all ages and expertise, whether they are children, adults or anatomists.

A view into the exhibition ANIMAL INSIDE OUT.

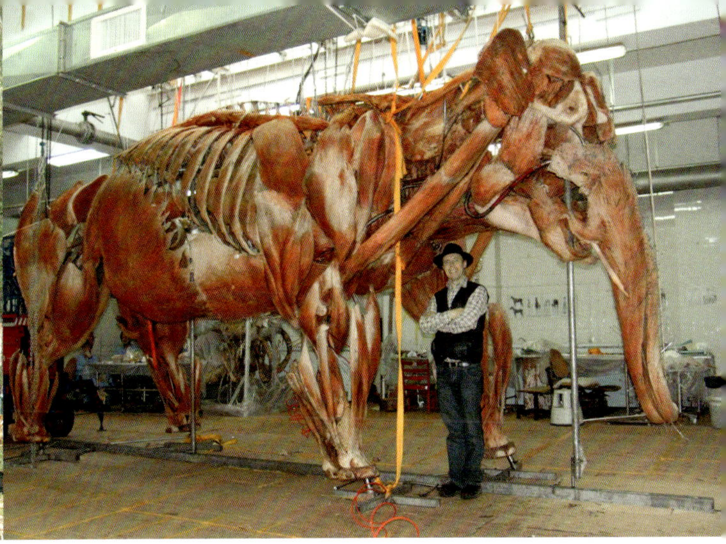

The female elephant "Samba" (right) died at the Neunkirchen Zoo in Germany. Finishing the monumental plastinate took three years.

History

For years now, I have been working on plastinating animals. A few years ago, I plastinated not only some smaller animals, but also big ones, such as a horse (2000), a camel and a gorilla (2003). Particularly these large animals require all of my creativity. The larger they are, the bigger the anatomical and technical challenge they present. I have even discovered a few new things when working on these giants, for example, on an elephant. This is because my team and I dissect these large animals in a detailed and careful manner that surpasses previous preservation techniques. In doing so, I feel like a researcher on an anatomical journey of discovery. For instance, we have been able to visualise that a giraffe's skin is more vascularised in its dark parts than in its light parts. This has never before been shown so clearly, as no one else previously has injected an entire giraffe with a contrast-enhancing resin that penetrates even the arteries of the skin, as we have done.

Over time we have created more and more interesting animal plastinates. Nevertheless, it was not until "Samba," the first elephant ever to be plastinated in the world that I thought of preparing an anatomical exhibition consisting solely of animals.

By preserving the elephant, I fulfilled the wishes of Dr. Norbert Fritsch, the director of the Neunkirchen Zoo in Germany. Samba died from cardiovascular weakness in February of 2005 and the giant body was donated to the Institute for Plastination, with the reciprocal agreement that the Neunkirchen Zoo would be the first to exhibit the plastinate. "Our usual method of disposing of deceased animal carcasses was either orderly disposal or cremation. We are pleased that their remains can be used to teach future generations about their anatomical splendour." said Dr. Fritsch. He later donated a second female elephant, "Chiana," after she was euthanized in April 2006 due to a broken leg and renal failure.

The idea for a first exhibition of animals rapidly took shape in my mind. Together with curator Dr. Angelina Whalley, ANIMAL INSIDE OUT was created. Never before have so many plastinated animals been shown in one exhibition, accompanied by informative facts specific to their anatomy. Most of the specimens, including the magnificent elephants and the blood vessel configuration of a horse's head, are being exhibited to the public for the first time. Only a few plastinates, such as *a gorilla, a giraffe, an ostrich,* and *The Rearing Horse with Rider,* were shown in previous BODY WORLDS exhibitions.

The majestic animals featured in ANIMAL INSIDE OUT were made available to us through cooperation with various university veterinary programs, zoos and animal groups. No animal was harmed or killed for the purpose of this exhibition. Among the animals in the exhibition, are human specimens, originating from the Institute for Plastination's body donation program. The generosity of these individual donors has made it possible to present human specimens in this and all of the BODY WORLDS exhibitions. Dr. Angelina Whalley and I, are honoured to be able to conserve and present these biological wonders of nature for anatomical study. We hope that this exhibition will show visitors the similarities between humans and animals, leading to a greater respect and appreciation for all animals.

The Making of Plastinates

Just like humans, most animals consist primarily of water. Not only is water essential for life, it is also an important part of decomposition. During the Plastination process, water in the body's tissue is replaced by reactive plastics, such as silicone rubber, epoxy resin or polyester resin, with a special vacuum technique. Body cells and the natural texture of surfaces are maintained in exactly the same state they were in before preservation – down to the microscopic level. Plastinates are dry and odourless, and thus particularly well suited for hands-on study. For the first time, authentic anatomical specimens can be preserved permanently in a lifelike and aesthetically pleasing manner.

These qualities make plastinates immensely valuable for research and the education of medical students and professionals, as well as laypersons. Books and images alone simply do not provide the same opportunity for visualizing the complex, three-dimensional structures of the musculoskeletal system, as wells as organs and their spatial relation to each other. Similarly, anatomical models, while three-dimensional, are merely schematic representations that do not show fine details. Nothing compares to learning from the real thing.

In addition to their educational value, plastinates are fascinating because they are authentic. Plastination halts decomposition in such a complete manner that observers do not perceive the body's interior as morbid or grotesque, and instead, its magnificence is revealed. In comparison to other preservation techniques there is no odour to interfere with visual perception. A beautiful plastinate, frozen between death and decay, facilitates a completely novel, captivating experience.

Specimens plastinated with silicone are cured with a special gas.

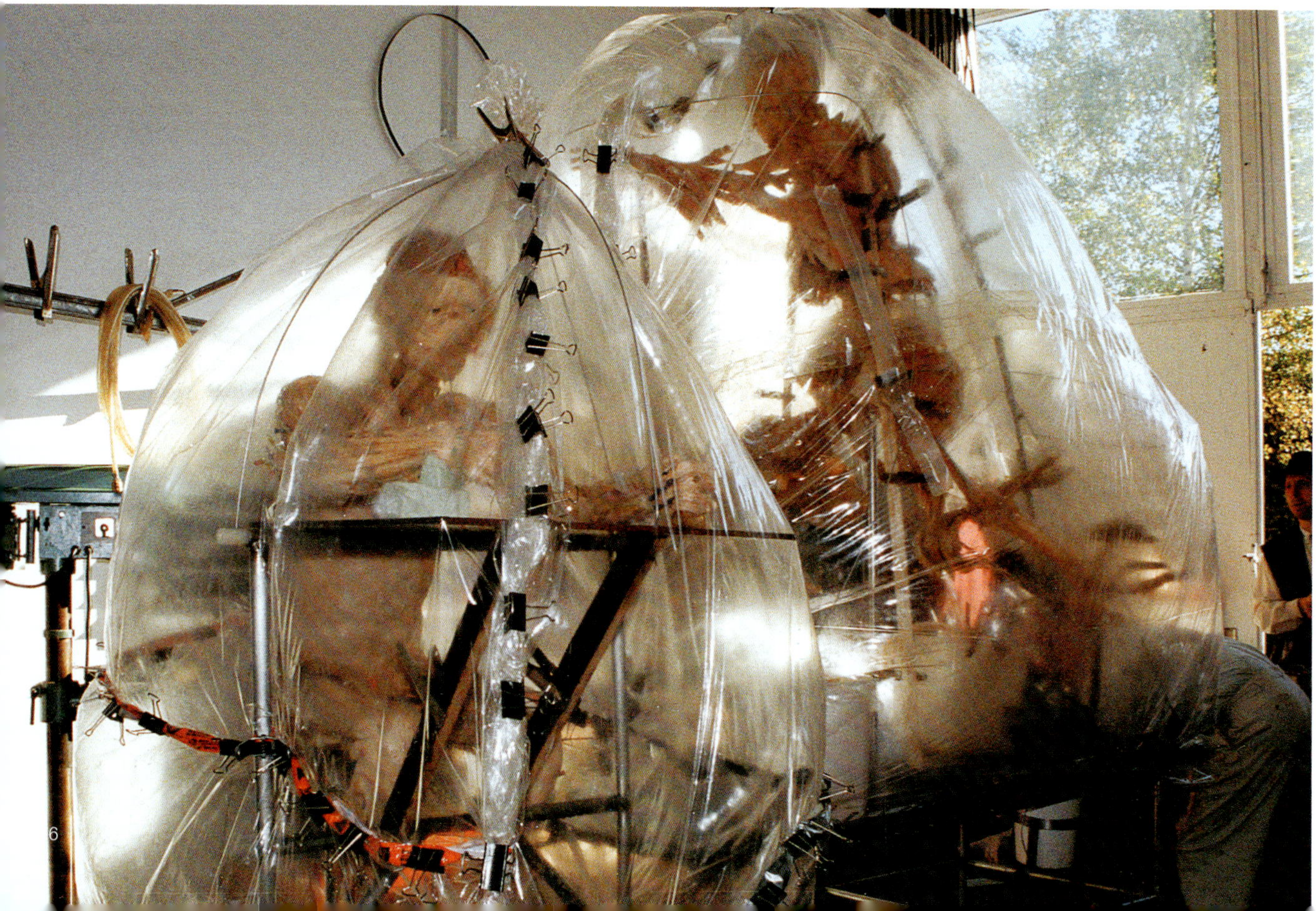

Forced impregnation. Acetone is squeezed out of the specimen and constantly sucked away.

Their life-like quality makes plastinates the visually most pleasing representation of permanently preserved animal specimens. This is particularly evident with the transparently plastinated body slices that reveal anatomical structures down to the microscopic level. The fact that plastinates will last literally forever justifies complex and detailed dissection procedures that previously would have been considered a waste of time. The detailed dissection and preparation of a human whole-body plastinate requires upwards of 1,500 working hours. In contrast, my team and I needed more than 8,000 hours, 8,000 litres (2,113 gal.) of acetone and 12,000 litres (3,170 gal.) of alcohol to prepare the horse plastinate with rider. The Plastination of the elephants, weighing 3.2 tons, required even more materials and labour. My team and I worked on "Samba", the elephant, for a total of 64,000 hours. About 4 tons (8,818 lbs) of silicone and 40,000 litres (10,566 gal.) of acetone were necessary to transform this giant land animal measuring 6 by 3.5 metres (20 by 11 ft).

Because Plastination hardens soft tissue like muscle and skin, new types of specimens become possible, such as transparent body slices that are three millimetres (0.1 inches) thick and reveal anatomical structures down to the microscopic level.

Plastination, like many revolutionary inventions, is simple in principle. After formalin is pumped into the body to stop decay, skin and fatty tissues are dissected and removed. Then all liquids and soluble fats are removed and replaced with acetone by means of diffusion. The acetone is then replaced with reactive plastics, specially developed for Plastination. This essential step, known as forced vacuum impregnation, is responsible for getting the liquid plastics into even the smallest cells of the specimen. After being immersed in a polymer solution the specimen is placed in a vacuum chamber. The negative pressure sucks the acetone out of the specimen, creating a volume deficit and inducing the plastics to stream into the tissues. This process utilizes the difference in vapour pressure between the volatile acetone and the polymer solution that has a high boiling point. Gradually, the specimen is infused with plastics. For thin body slices, this process can be completed within a few days. For whole bodies, it may take several weeks. The specimen is removed from the liquid polymer bath only when the vacuum has dropped to less than one hundredth of normal atmospheric pressure (<5mmHg) and there are only a few more acetone bubbles slowly escaping from the specimen. After removal from the bath, the specimen is positioned as desired and then hardened or cured with gas, light or heat.

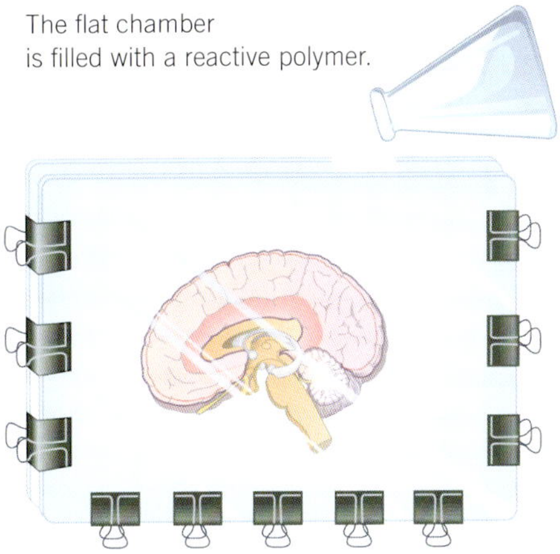

Cross-sectional Plastination. Tissue slices that have been saturated with a reactive polymer are placed in a flat container where additional polymer is poured over them.

PLASTINATION PROCESS

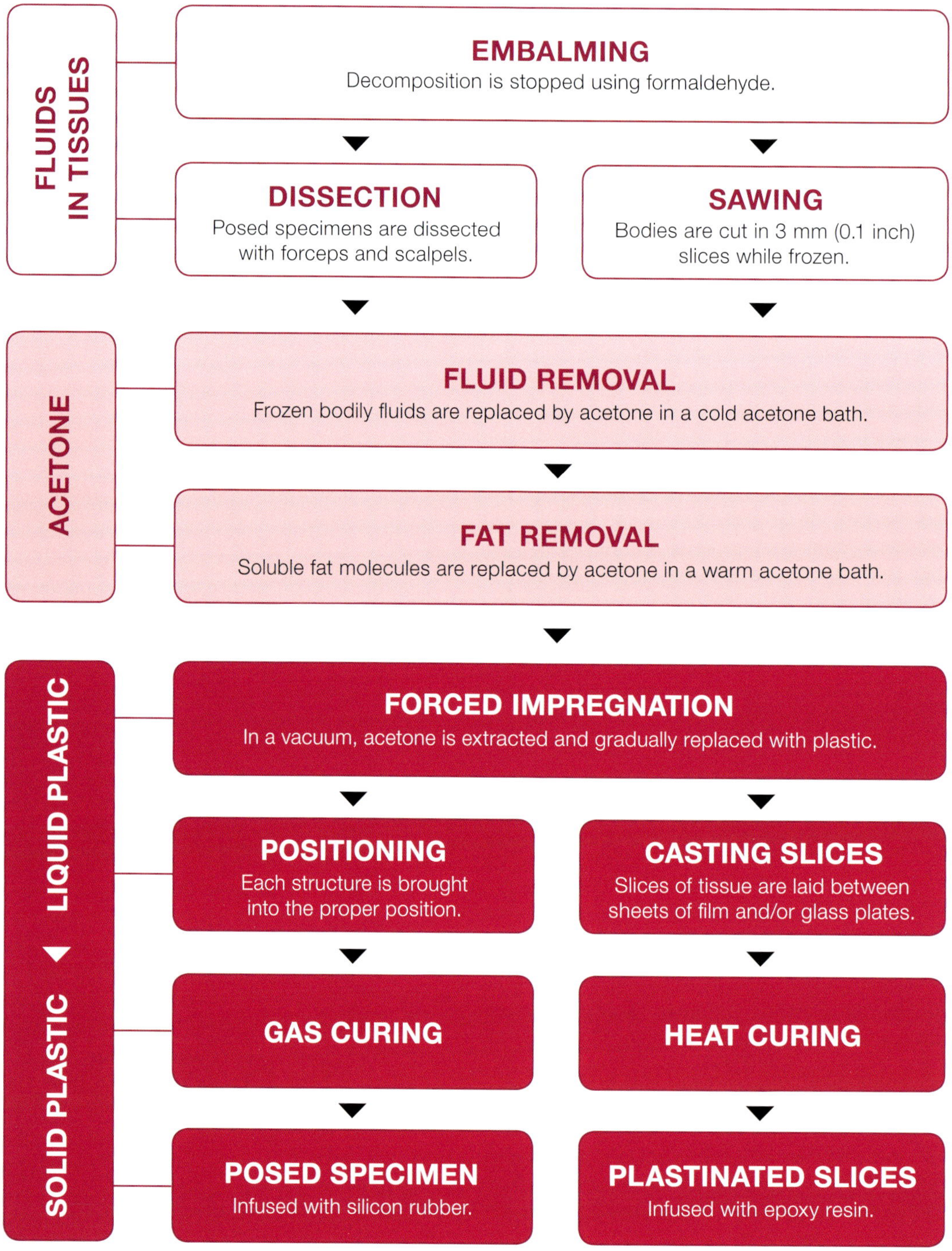

Overview of Silicone Plastination

1. Embalming and Anatomical Dissection

The first step of the process involves halting decay by pumping formalin into the body through the arteries. Formalin kills all bacteria and chemically stops the decay of tissue. Using dissection tools, the skin, fatty and connective tissues are removed in order to prepare the individual anatomical structures.

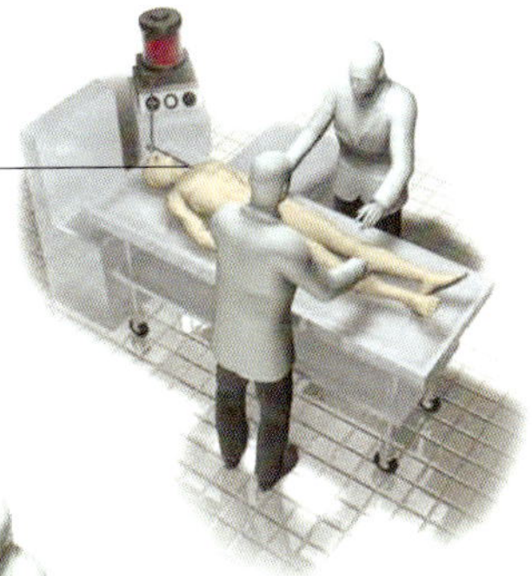

2. Removal of Body Fat and Water

In the first step, water and soluble fats are dissolved from the body by placing it into a solvent bath (e.g. an acetone bath).

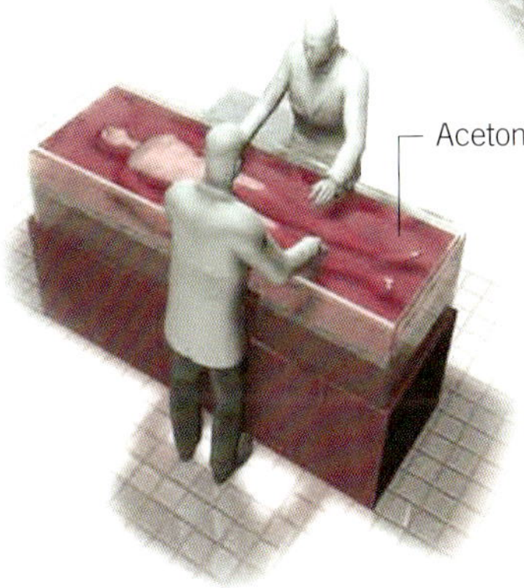

3. Forced Impregnation

This second exchange process is the central step in Plastination. During forced impregnation a reactive polymer, such as silicone rubber, replaces the acetone. To achieve this, the specimen is immersed in a polymer solution and placed in vacuum chamber. The vacuum removes the acetone from the specimen and helps the polymer to penetrate every last cell.

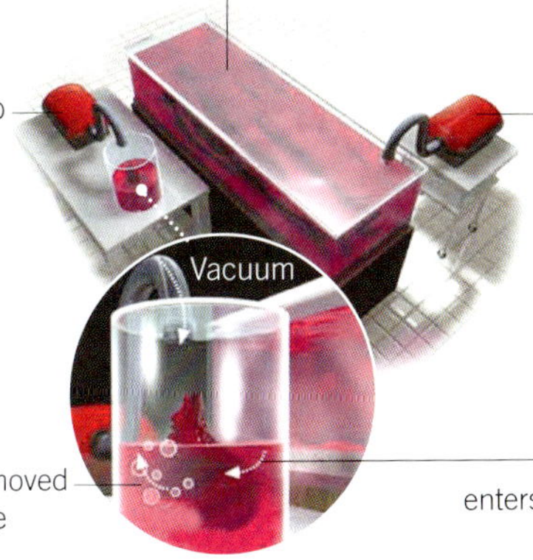

4. Positioning

After vacuum impregnation, the body is positioned as desired. Every single anatomical structure is properly aligned and fixed with the help of wires, needles, clamps and foam blocks.

Positioning

5. Curing (Hardening)

In the final step, the specimen is hardened. Depending on the polymer used, this is done with gas, light or heat.

The dissection and Plastination of an entire body requires about 1,500 working hours and normally takes about one year to complete.

Images courtesy of the DENVER POST

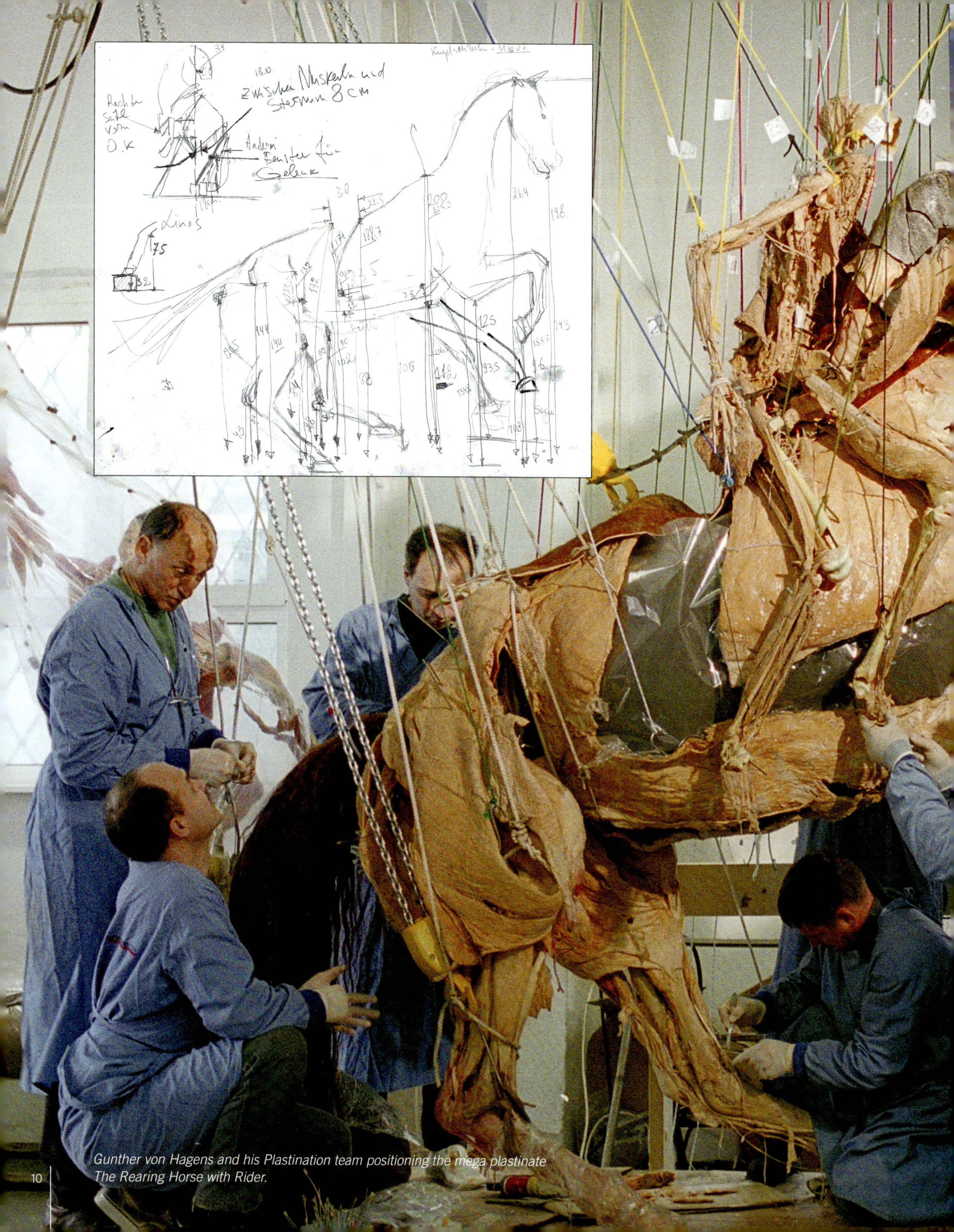

Gunther von Hagens and his Plastination team positioning the mega plastinate The Rearing Horse with Rider.

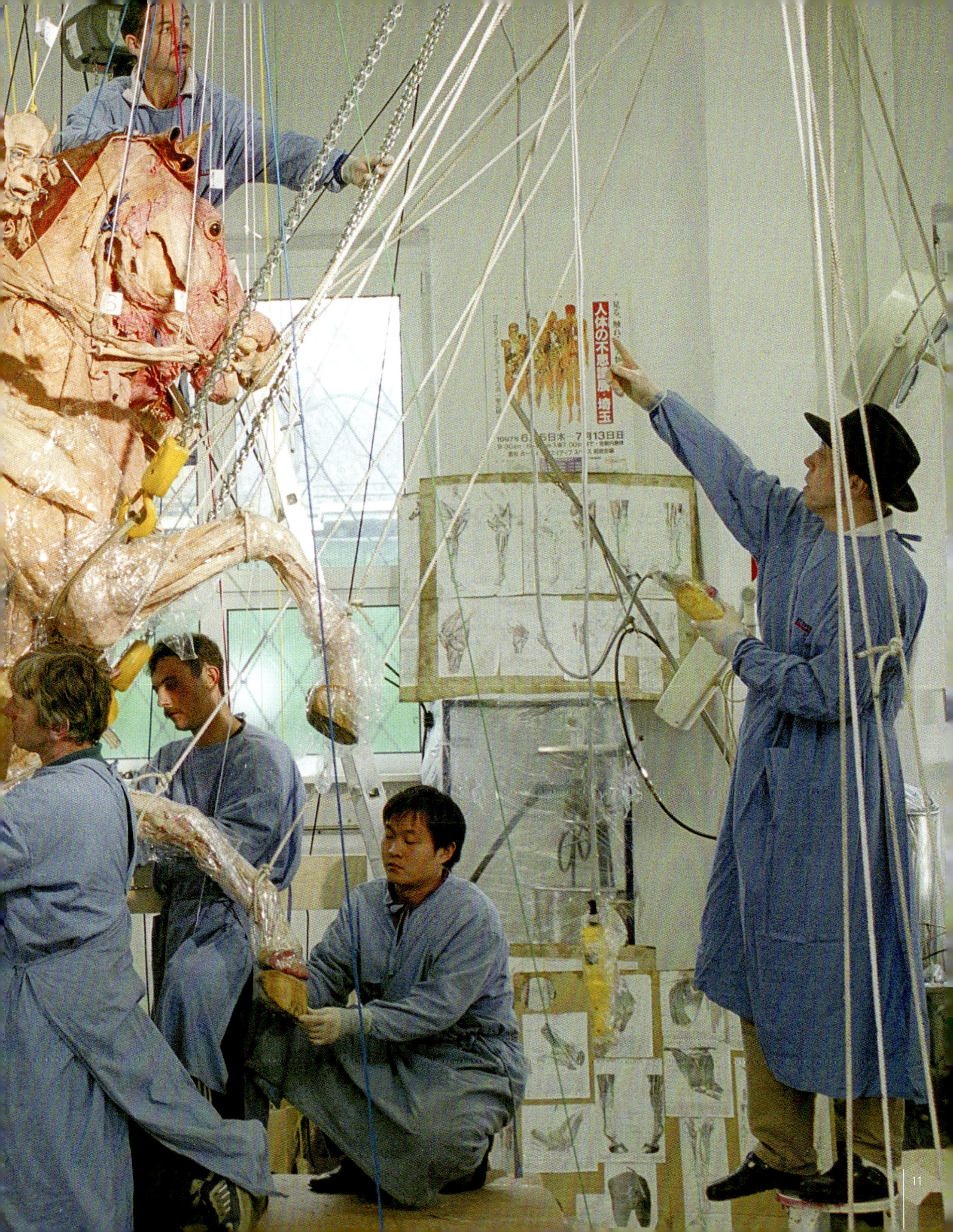

The Idea Behind Plastination

I invented the Plastination process in 1977, when I was still at the Anatomical Institute of Heidelberg University. It was patented in the years 1977–1982, and I have continuously developed it ever since. The origins were less than spectacular. While working as an assistant researcher at the Anatomical Institute for the first time, I saw specimens embedded in plastic blocks and wondered why the plastic encased the specimens rather than being injected into the specimens which would have rendered them stable. It was a question that would continue to haunt me.

Some weeks later, I was asked to produce serial cuts of human kidneys for a research project. As I needed only every 50th slice, it seemed to me that the usual process of embedding the kidneys in paraffin and subsequent cutting into the finest of serial slices was rather cumbersome. While shopping at a small grocery store on campus and watching the sales woman cut ham, I had an idea: I ought to use the same type of cutting machine for slicing kidneys.

That is how a "rotary slicer," as I referred to it in my grant application, became the first investment in Plastination. I used liquid Plexiglas to embed the kidney slices. The air bubbles introduced while stirring the hardener into the solution had to be extracted under vacuum conditions. While watching these bubbles, the crucial idea came to me: under vacuum conditions, it should be possible to impregnate a piece of kidney soaked in acetone with plastic by extracting the acetone as bubbles, just as the air was extracted in this process. Numerous acetone bubbles escaped from the specimen when I tested the hypothesis, but after one hour, the kidney specimen was coal black and had shrunk.

Most people would likely have given up at this point; however, I had enough of a background in physics and chemistry to understand that the black colour was due to the refraction index of the Plexiglas, and that the shrinking was due to an overly rapid impregnation process. Thus, a week later I repeated my experiment using liquid silicone rubber. I used a slow impregnation progress, going through three sequential silicone baths in order to prevent premature hardening of the silicone including the specimens. After hardening my specimen in an oven, I held in my hand the first presentable plastinate.

The very first plastinates were everything but attractive. Surplus resins obscured their surfaces. Transparent tissue slices contained numerous air bubbles. Frequently, the specimens were ruined when the plastic solution polymerised prematurely. It took me a whole series of additional inventions to develop Plastination over the next three decades, from a hardly viable idea into a widely recognised preservation technique.

Anatomy Art

ANIMAL INSIDE OUT presents biological facts with visually stimulating experiences. While the aesthetic appearance and effect of the specimens is in the eye of the beholder, no one completely escapes their fascination. Just as most people appreciate the colours and shapes of flowers, rather than their function, people are most often impressed by the appearance of an animal plastinate, more than the function of its anatomy. Many people simply want to see the enormous range of shapes and sizes of animal organs. Viewing these, however, will naturally lead to questions about their functions.

When looking at the plastinates, and in particular the whole-body specimens, both feelings and intellect are involved. The emotional response is direct; it happens without any awareness. The beautiful, active poses of the plastinates prevent any sense of revulsion that could occur from looking at the dead. It is intentional that the specimens are not accompanied by an overwhelming amount of information, because it encourages a direct connection between the observer and the animal. The exhibition presents an engaging experience for all and is very well suited for those who want direct and immediate comprehension of anatomy without too many wordy explanations. It is absolutely educational and fascinating to simply look inside an animal's body and learn through one's own eyes.

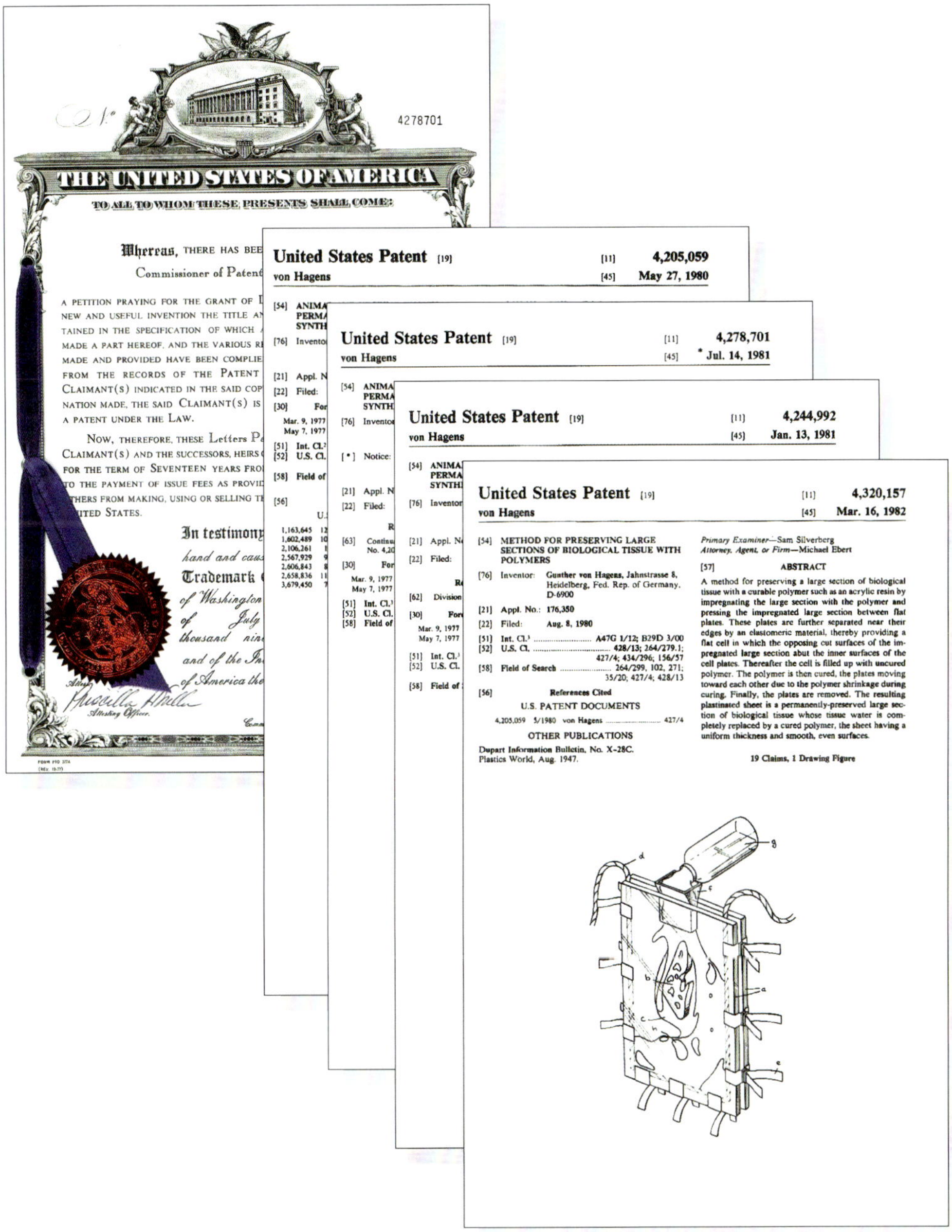

Patent Rights

The Plastination process is protected by a series of patents, particularly in the United States. Plastinating specimens for noncommercial purposes such as medical instruction or museum exhibits, however, is not subject to any such limitations.

While plastinates are not intended to meet any artistic requirements and I myself do not attempt to create art with my plastinates, they may still fulfil artistic criteria. Just like works of art, attractive plastinates have an aesthetic effect and evoke emotional responses.

To avoid any misunderstandings, I have coined the term "anatomy art." To me, this means the "aesthetic and instructive representation of the body's interior." Representation in this sense encompasses both presentation and an artistic quality.

Anatomy art in this sense continues a tradition from the Renaissance, when art was considered a mastered craft. It is further developed in Plastination, as this preservation method, for the first time, allows soft tissues such as muscles or fatty tissue to be hardened at will. This facilitates entirely novel types of specimens, including 3-mm-thin body slices that are as transparent and colourful as stained glass. On the other hand, the hardening of soft tissues allows an aesthetically pleasing and educational fragmentation of the body. This fragmentation in turn facilitates a rearrangement and opening up of the body that reveals entirely new views of its interior.

In addition to these dissection techniques, the authenticity of the specimens also contributes to the fascination caused by the exhibition. Especially in today's media-based world in which citizens increasingly depend on indirect information, individuals still sense that any copy has been mentally "pre-processed", and therefore, always constitutes an interpretation. In this sense, the ANIMAL INSIDE OUT exhibition satisfies an existing desire for unadulterated originality.

Discovery in Anatomy

Visiting ANIMAL INSIDE OUT is similar to a journey that serves the purpose of watching animals. In this sense, it is a type of safari. In this safari, each animal is studied for its own merit and, at the same time, comparisons between different animals are facilitated.

The animal kingdom is generally classified into phyla, classes, orders, families, genera and species. As trite as it sounds, it is still remarkable: the body's segmentation into head, torso and extremities is characteristic for all vertebrates. Another characteristic principle of all vertebrates is that of polarity: at one of the poles, there is an opening for the uptake of food, at the opposite pole there is the outlet of the digestive system. In addition, all vertebrates are bilaterally symmetrical to a large degree: a longitudinal cut through the body will split it into two halves that mirror each other and are almost identical in their contours and symmetries.

The bones of the skeleton are connected to each other in either a mobile or an immobile fashion. Joints serve to freely move two or more bones against each other.

The complexity of the animal body is particularly impressive in the interactions of the muscles. In general, three types of muscle are distinguished: smooth muscle, striated muscle and heart muscle. Every muscle is an organ made up of various elements. Skeletal muscles are made from striated muscle tissue. When stimulated by the nervous system, a muscle will exert a pulling force on the bone to which it is connected by tendons. Even when relaxed, the muscle has a certain tension, called tone, that is dependent on the nervous system.

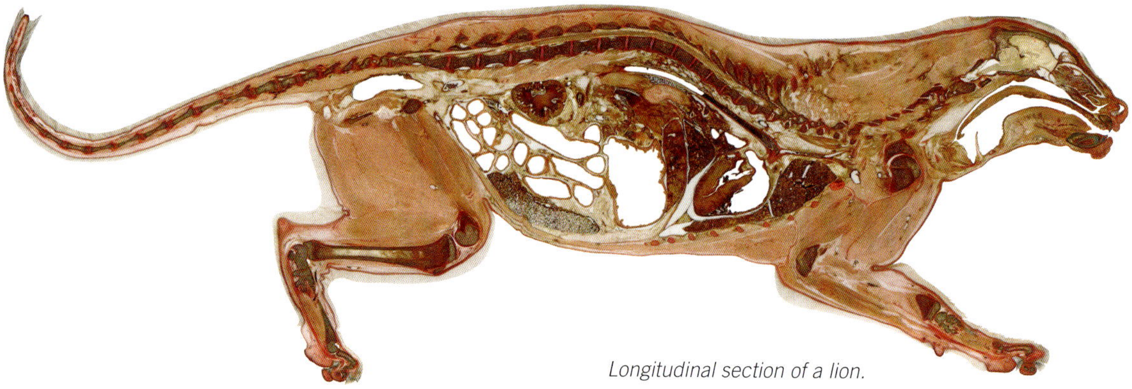

Longitudinal section of a lion.

Striated muscles serve various functions. They aid the circulation of the blood and generate warmth inside the body. They also support the chest and pelvis, carrying the internal organs. More importantly, they enable the movement of head, extremities, and torso. These muscles can be stimulated by the nervous system to bend, twist, lift and lower.

Striated muscles are found mostly in the chest, abdominal walls and limbs. They are controlled by the living being itself. Smooth muscle and the heart muscle, in contrast, work almost independently of the being they belong to. They are responsible for the internal processes in an organism, such as digestion. The bronchi, the midgut, and the rectum contain smooth muscle.

In the exhibition, a remarkable number of similarities between organisms can be observed. This includes the seven cervical vertebrae of humans and horses that are also found in almost all other mammals – even giraffes. There are some exceptions; for example, two-toed sloths have only six cervical vertebrae and three-toed sloths have nine.

It is true in general, that many or all members of an animal phylum will have similar organs. The most prominent example of this are the front extremities of vertebrates: the pectoral fins of fishes, the wings of birds, the arms of humans, and the front legs of amphibians, reptiles and mammals, are phylogenetically related. This type of conformity is referred to as *homology*.

Analogy, in contrast, refers to organs that fulfil the same function but are not phylogenetically related. Instead of being derived from a common ancestor, analogous organs evolved independently because they live in a similar habitat and share a common way of life. Examples of analogous organs include the lungs of mammals, the gills of fish, and the wings of insects and birds.

Convergent evolution is a term that refers to a similarity of some organs in beings that are not phylogenetically related. These similarities usually have developed because the different groups of animals have been exposed to the same or similar environmental conditions. For example, sharks, fishes, and dolphins have a streamlined body with a fin on their backs.

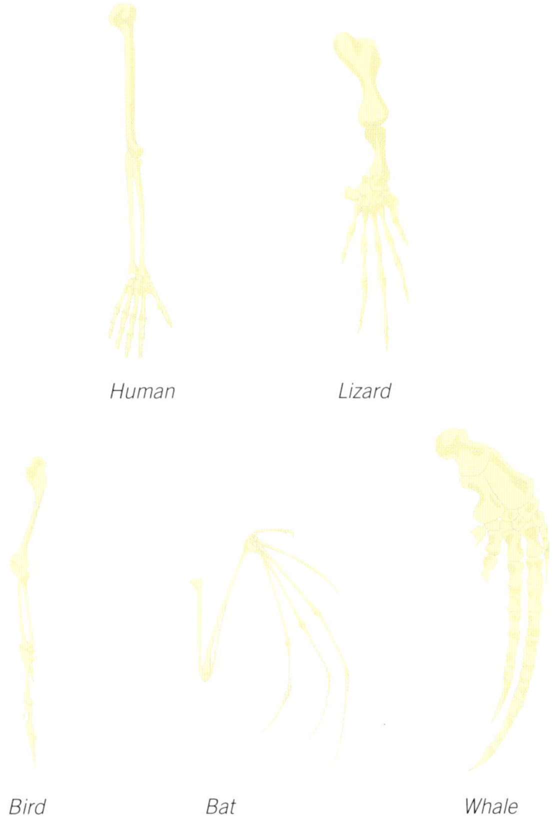

The front limbs of many animals are very similar because they have a common evolutionary origin. This similarity is referred to as homology.

All of this means that not everything that looks the same is actually the same.

There is much to discover in ANIMAL INSIDE OUT. This introduction highlights just a few examples of the incredible diversity as well as commonalities within the animal kingdom. More can be found by carefully comparing the plastinates in the exhibition and this catalogue. Through the exhibition, we aim to encourage an in-depth study of animals and improve understanding of the anatomical relationships between form and function. At the same time, ANIMAL INSIDE OUT enhances our ability to discover the extraordinary in the ordinary.

Big Animal, Big Effect

It is not surprising that the plastinates of large animals amaze us, given how humans in general are fascinated by everything big. The aesthetic impression of anything is not only determined by its shape, but also by its size. Often, the sheer mass of something huge will inspire quiet awe and surprised amazement simply because it exceeds all average measures. On the other hand, delicate shapes such as plastinates of dissected nerves or vessel configurations of thousands of tiny arteries, forming a complex network, may enrapture us as well.

If we think about it, being great is not a characteristic of any thing itself, but a feature that we assign it. The vast ocean, high mountains and the starry sky are not great of their own accord, just as they are not beautiful or ugly of their own accord. All of these attributes are simply assigned by our ideas and expectations, which in turn depend on cultural norms. Whether or not we think of something as great does not solely depend on its measurable size. As Immanuel Kant, philosopher of the Enlightenment period, emphasised, the great is deemed so only in comparison. There is only the great, the large, and the big, because there is also the small. The smaller and more powerless we humans feel compared to what we see, the greater and more powerful the object viewed appears to us.

Monuments such as the statues of Pharaoh Ramses II at Abu Simbel in Egypt impress us, because we seem like nothing next to them, due to our physical and historical dimensions.

Natural phenomena such as giant mountains or boundless oceans, and cultural achievements such as the Asian statues of Buddha, the pyramids of Egypt, or the skyline of Manhattan, appear as great as they do because they stretch our senses to their limits. Large animals in the exhibition may have a similar effect, as the visitor feels small and powerless next to them.

However, they also appear great to us because they transcend what we consider normal, what we are used to and what we experience every day. They reveal what we had not expected; they make visible what we could not imagine before. This experience is similar to that of a cave dweller who for the first time steps onto the surface of the earth, seeing sunlight for the first time, or that of someone living inland who for the first time sees high mountains and the vast sea: they are amazed and their quantitative perception of extraordinary size is, almost automatically, accompanied by a qualitative judgment of greatness. Visitors who come unprepared into the exhibition will have a similar experience. They will be utterly amazed by the variety of shapes and the sheer size of the large animal plastinates.

Vulnerable Nature

Anatomical education and aesthetic fascination are not the sole purposes of the exhibition. It also aims to draw attention to the vulnerability of the animal world and strives to instil a greater appreciation for endangered species. We cannot expect humans to treat animals with care and respect unless they thoroughly understand the wonders and unique characteristics of their nature. However, idealizing the animal world is not adequate either. While animals can be loveable and good, they can also possess destructive powers.

Nature does not assume any responsibility for the preservation of life. Cats will not stop catching mice. Lions will hunt and devour antelopes, completely unaffected by their need for survival. In addition, nature has been quite wasteful throughout history when it comes to creating and destroying its elements. It is undoubtedly a shame that roughly 16,000 species of plants and animals are currently threatened, endangered or critically endangered. Partly due to human recklessness, they may disappear forever from our globe. However, more than 99% of all plant and animal species that have existed on Earth have become extinct through the process of evolution, independent of human activities.

Animals have the right to defend themselves and prevent humans from forcefully destroying their species. Still, humans cannot be the guardian of all forms of life. We cannot keep the lion from killing the antelope, and we cannot defend the mouse's interests against those of the cat. In other words, we cannot assume responsibility for the animal world itself, but only for the consequences of what we do to it. We cannot protect it from itself, but only from ourselves. Just as we have to continue to protect ourselves from the animal world as well.

We are chiefly interested in the protection of mammals. They are close to us, both in an evolutionary sense and because we interact with them on a daily basis. We also care for fish and birds. But what about pests, in particular those that occur in large numbers? There are ordinances requiring the destruction of pests. However, even without legal requirements, humans will fight certain animals, such as mice, and in some cases, this has nearly resulted in the eradication or decimation of a species, such as the May bug in Europe. What about all those creatures that belong to classes other than mammals and birds, for example insects, worms, and molluscs? Unlimited protection of animals would also be opposed to the killing of a bothersome fly.

Yet, the exhibition aims to convey more respect for the animal world by making visible the external relationships and astounding complexity of all animal forms. Animals, like humans, are capable of suffering. In addition, beings capable of feeling pain are, in general, not interested in suffering. This need is of the utmost moral importance.

In human history, animals have at times been viewed as inferior beings, subject to our whims. Roman law and philosophers of modern times, including René Descartes and Immanuel Kant, provide examples of this view. Descartes even likened animals to machines, absent of feeling, evoking protest from Leibniz and Voltaire who emphasized that animals indeed have this capacity. Some philosophers agreed with this perspective but argued that this alone did not confer a particular right to careful treatment. Augustine of Hippo, for example, did not deny that animals were capable of sensations. But he still thought that humans were allowed to do with them whatever they pleased, because animals were without reason. Voltaire and Arthur Schopenhauer disagreed. They supported a similar position as Jeremy Bentham, who argued that a being's ability to think or to speak was less important than whether it was capable of suffering. Rousseau thought along the same lines when he wrote: "If I am obliged not to cause suffering for a fellow human being, so it appears that this is less so because he is a reasonable being than because he is a feeling being: This is a characteristic that animals and humans share, and it has to give one the right to not be uselessly mistreated by the other, to say the least."

Now that we posses a better understanding of physiology, we know that all animals with a central nervous system feel pain – be they fish, amphibians, reptiles, birds or mammals. However, to what degree animals with a ventral cord are capable of suffering, such as insects, crustaceans and worms, or even simple invertebrate coelenterates with a nervous system, such as polyps, jellyfish, and corals, is a question that is much harder to answer.

Thomas More and Immanuel Kant, even in their days, were of the opinion that treating animals in a violent and cruel way might dull our senses so we no longer feel empathy for other suffering beings, causing us to act more brutally in general.

In this context, the question of keeping domestic animals in as natural an environment as possible can hardly be avoided. Any mistreatment of animals draws attention to their vulnerability that in itself offers an ideologically neutral reason for treating them with respect. Animal abuse, painful killings and confined living conditions for mass breeding – chickens in battery cages, cattle in stalls and chains – are nothing less than undignified.

In addition, one has to ask whether the painless killing of chickens, pigs, lambs, and calves can be morally justified. Often, slaughter is considered less ethically controversial than animal abuse. Normally, it is justified by stating that if no one were to eat cows, chickens, pigs, lambs, and calves would not even exist, as they are raised almost exclusively to provide food. In addition, it is argued that because animals are incapable of thinking about the future, they only feel joy and pain in the present moment; therefore, abusing animals in experiments or industrial mass breeding is not acceptable, but painless killing is. However, there is controversy over whether or not higher order mammals such as whales, dolphins, dogs, cats, pigs, and calves actually have a sense of a future.

Cooped up animals, dead eyes – everything for the well-being of humans? Experimental rabbits in a cage.

ANIMAL INSIDE OUT cannot provide definitive answers to these difficult ethical questions. However, the exhibition may cause you to think about them intensely.

Living Beings like Us

As perplexing as it may seem, in terms of scientific practice, there is hardly any connection between human anatomy and animal anatomy. Veterinarians deal solely with animals; physicians treat only humans. There are several reasons why veterinary anatomy textbooks only touch on human anatomy and why animal anatomy plays almost no role at all in human anatomical atlases. One is the simple fact that animal and human medicine have almost no interface and are strictly separated at universities. Additionally, this strict separation mirrors the traditional chasm between humans and the animal world.

Although it was shown a long time ago that there is an evolutionary connection between humans and all other living beings, and numerous similarities are known to exist between humans and all other vertebrates, especially mammals, many people perpetuate the age-old barrier between humans and animals. In some languages, this division is reflected in how different verbs are used to describe identical actions of animals and humans. For example, animals are said to feed while humans eat. Animals are suckled, and human babies are breastfed. Because of our egocentric tendencies and ability to speak and think abstractly at high levels, many of us believe humans are more valuable than other creatures, even when they die, and even when only their dead bodies remain. This is also reflected in some languages; for instance, different terms are used for the dead bodies of humans and animals, distinguishing between carcass and corpse. In cases where this distinction is made, the animal-related terms can sometimes be used for humans, but usually with a derogatory, demeaning implication. Acting like an animal means acting without manners, being rough and inconsiderate.

However, there is no justification for this language-based arrogance toward animals. Not only are animals capable of enduring far more stressful conditions than humans – be it cold, heat, or rain – many of them are also faster and stronger than we are. Almost every human organ finds more than its match in the organs of animals. For example, the kidneys of a desert mouse work much more effectively than ours. They can survive better in a dry environment because they excrete less water with its nitrogen compounds and urea. Night moths have a range of hearing between 40,000 Hertz and 80,000 Hertz, while human hearing can detect frequencies between 16 Hertz and 16,000 Hertz. Horses, adapted to an optimum flight reaction, also have an extremely good sense of hearing. Dogs have a much better sense of smell than humans do. The cerebellum of birds, which controls coordination and balance during flight, is relatively more developed than that of humans. Only the human cerebrum, which contributes to higher mental functions like thought, emotion and reason, is not surpassed by that of any animal.

The traditional idea that humans are unique is exaggerated, as is the strict separation between humans and animals. Humans, also known to scientists as *Homo sapiens sapiens*, belong to the order of primates and the genus *Homo*. This species differs from other hominids, but not substantially. The biochemical discovery that human DNA differs from that of chimpanzees by only 1.6% illustrates this perfectly. We share almost 99% of our genes with large apes (chimpanzees, gorillas and orangutans). This lack of distinction is further supported by the fact that there are no fundamental anatomical or physiological differences between our brain and those of other vertebrates.

To quote Michel de Montaigne: "We are neither above nor below the rest. 'All who are under the sky,' says the sage, 'experience a like law and fate.' There is some difference, there are orders and degrees, but they are under the aspect of one and the same nature. [...] That which he confers upon himself in his own opinion and fancy has neither cause nor substance."

However, integrating humans into the animal kingdom does not necessarily need to be viewed as lowering humans to the level of animals. Just as well, it might be seen as an uplifting or upvaluation of animals. For if humans are related to animals, animals are also related to humans. In this sense, this exhibition intends to return some of the respect and appreciation that humans have taken away from animals for centuries, in order to claim them solely for ourselves.

Human evolution: from about 160,000 years ago until today. Humans belong to the family of apes.

Dr. Gunther von Hagens began his medical studies at the University of Jena in 1965. He was arrested while attempting to flee East Germany after distributing leaflets in protest against the invasion of Czechoslovakia by Warsaw Pact troops. In 1970, West German authorities secured his freedom. He completed his medical studies at the University of Lubeck in 1973 and a year later he received his license to practice medicine. He then moved to the University of Heidelberg, where he completed his doctorate in the Department of Anaesthesia and Emergency Medicine. In 1975, he began to work at the Institutes of Anatomy and Pathology, which ignited his passion for invention and to teach anatomy in a new way.

It was in 1977 at the University of Heidelberg that he invented Plastination, a vacuum technique to permanently preserve anatomical specimens with reactive plastics specially developed for this purpose. He subsequently founded BIODUR® Products to market the respective polymers and equipment to other institutions around the world. In order to bring Plastination to a wider audience, he founded the Institute for Plastination in 1993. Soon after, together with his colleague and wife Dr. Angelina Whalley, he created the first anatomical exhibition of real human bodies, titled BODY WORLDS, which has now seen by more than 45 million people worldwide. They later launched ANIMAL INSIDE OUT, the first exhibition dedicated to plastinated animals, in 2010.

His popular success in Plastination has been accompanied by academic accolades. In 1996, he became a visiting professor at the School of Medicine in Dalian. That same year he became Director of the Plastination Centre at the State Medical Academy in Bishkek, Kyrgyzstan, where he was awarded the title of Honorary Professor. Since 2004, he has been a visiting professor at New York University College of Dentistry.

In 2006, he established Gubener Plastinate GmbH, a Plastination laboratory in Guben, Germany. It is considered the world's leading Plastination research, development and preparation facility. The campus also houses the PLASTINARIUM, which promotes public anatomy by granting access for all visitors to observe the preparation of permanently preserved, anatomical human and animal specimens.

The international Association of Science-Technology Centers (ASTC) honoured him in 2013 with a lifetime achievement award for outstanding contribution to the public understanding of science.

Angelina Whalley

ANIMAL Anatomy

The essential requirements in the life of an animal are quite simple. It must find food and convert it into energy, defend or protect itself, reproduce and colonise new territories.

The methods with which different animals meet these requirements within their special habitats are, however, extremely diverse and complex, and are determined by the anatomy of the animals.

In the following chapters, the major systems of the body and their modes of operation are described.

The specimens depicted in this catalogue are unique. The exhibition ANIMAL INSIDE OUT includes a large collection of these animal specimens and certain BODY WORLDS exhibitions show some animal plastinates. Due to the individual nature of the specimens and certain space limitations not all animals in this catalogue are shown in any one exhibition.

Dr. Angelina Whalley started her medical career at Freie Universität in Berlin and completed her doctorate with an experimental dissertation on kidney physiology at the University of Heidelberg in 1986. The same year, she obtained her license to practice medicine. For several years, Dr. Whalley did scientific research at the Institute of Anatomy as well as at the Institute of Pathology for the University of Heidelberg.
Dr. Whalley has been the creative and conceptual designer of all BODY WORLDS exhibitions since their debut in 1995. As exhibition curator, she introduced the concept of comparative anatomy in the exhibitions, and is responsible for presenting BODY WORLDS in a meaningful way that optimizes the aesthetic and educational value of Plastination. Her memorable, thematic story-telling as well as her accessible teaching approach has inspired more than 45 million people around the globe to discover the wonders of the human body and adopt healthier lifestyles along the way.
In ANIMAL INSIDE OUT, the first exhibition of plastinated animals, she created a symbiosis of anatomy and didactics, in a captivating presentation, aiming at increasing awareness for the wonders of nature and the endangered biodiversity on our planet.
Since 1997, Dr. Whalley has also been the Director of the Institute for Plastination in Heidelberg, Germany and Director of BIODUR® Products, a company that markets Plastination formulas and auxiliaries to more than 400 medical schools and universities worldwide. In 2003, she founded Arts & Sciences Exhibitions and Publishing GmbH, a company planning and touring scientific exhibitions worldwide as well as publishing related materials. She is the author of several best-selling books and catalogues that chronicle the exhibitions and is producer of a related documentary.

The 200 bones of a lion's skeleton form a smooth inner scaffolding to which more than 600 muscles are attached. All cats have a similar structure. In fact, even by looking at the skeleton (excluding the head), experts usually cannot identify which big cat species it belongs to.

Strong Support

The bodies of most vertebrates have a stable framework, the skeleton, which gives them their basic shape and posture. Attached to the skeleton are muscles. Together the bones and muscles form a system of levers, transforming muscle action into movement.

The skeleton is the most diverse organ system across the various groups of animals. It is adapted perfectly to the different habitats of each animal.

The biggest and best known animals of the world are the vertebrates, animals that have a spinal column. They have an internal skeleton, called an endoskeleton, that consists mostly of hard bone or cartilage.

The skeletons of vertebrates consist of many individual bones that are connected by joints. Strong ligaments hold the ends of the bones together.

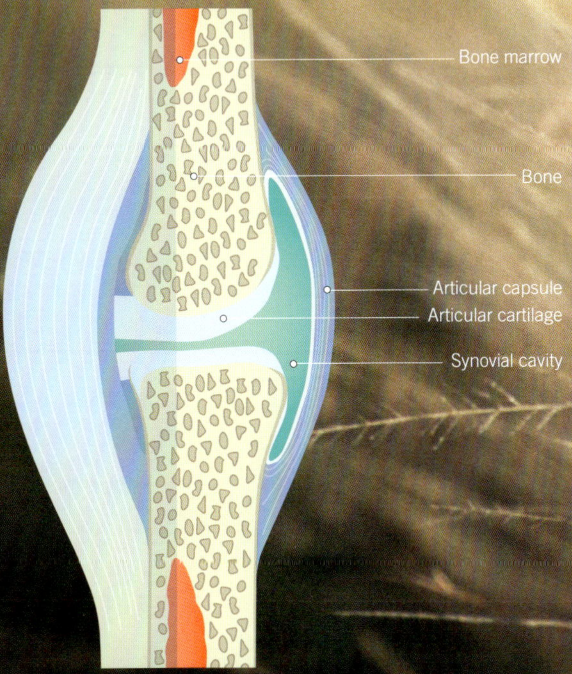

Typical joint structure of a vertebrate

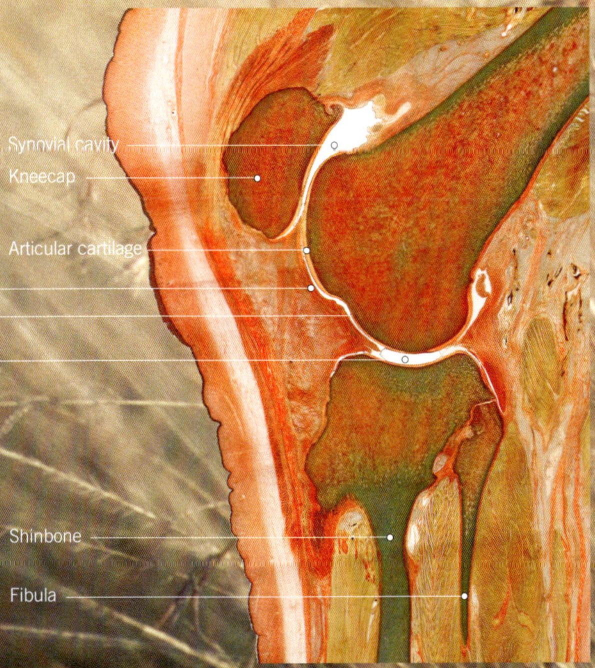

Longitudinal section through the knee joint of an Asian elephant leg
(Elephas maximus indicus)

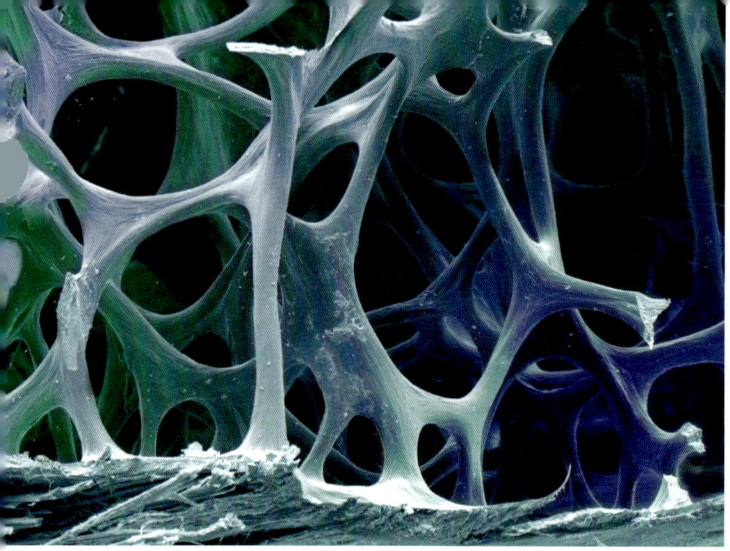

Trabeculae in 15-times magnification show how cancellous bone adapts to static requirements and, depending on the mechanical stress, forms lines of tension.
This design gives the bone great stability despite its light weight.

Bones have a special structure:
they have a compact, hard outer layer called cortical bone that surrounds a sponge-like tissue made up of *trabeculae*, called cancellous bone.

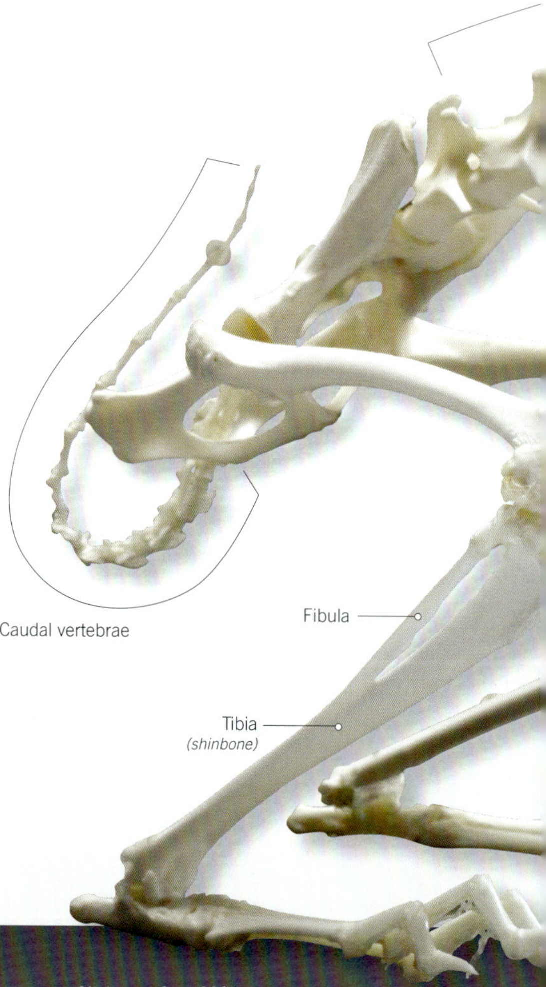

Skeleton of a rabbit
(Oryctolagus cuniculus forma domestica)

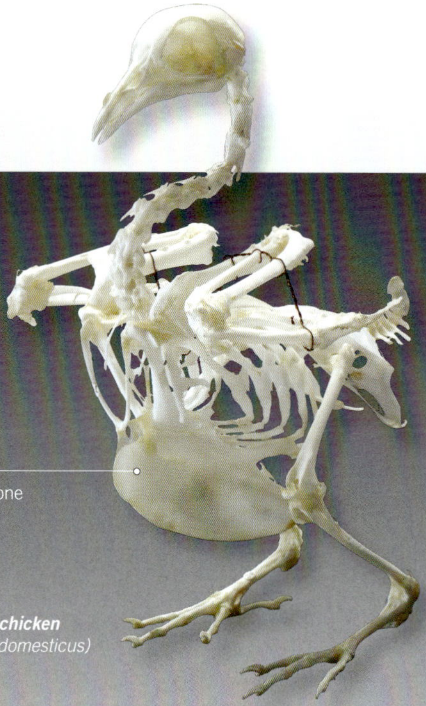

Skeleton of a chicken
(Gallus gallus domesticus)

In chickens, the skeleton primarily serves as a supporting structure, allowing them to walk on two legs.
As in all birds, some of their bones are filled with air, keeping their weight as low as possible.
Birds usually have a bony protrusion on their breastbones, the keel, where the flight muscles attach.

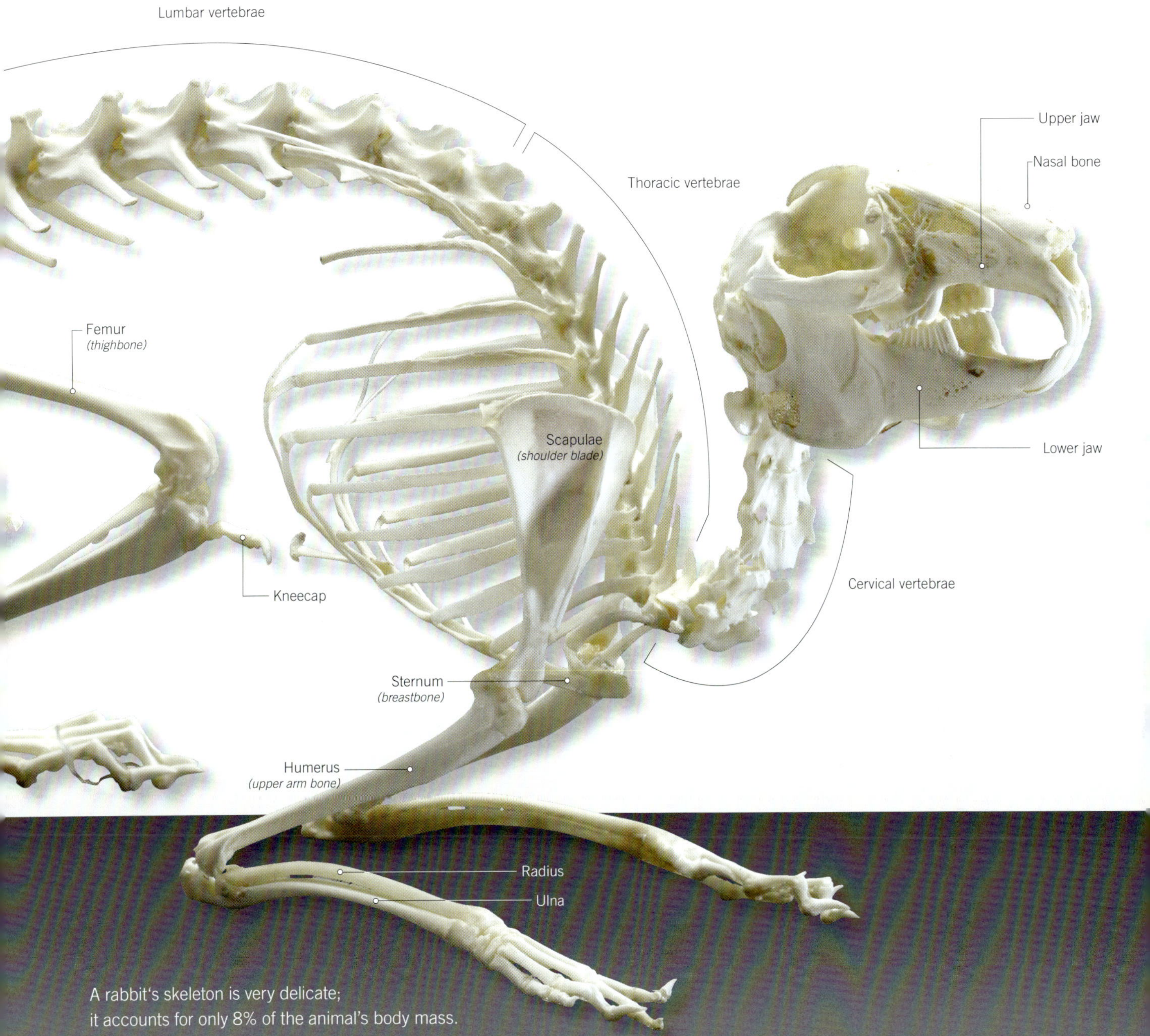

A rabbit's skeleton is very delicate;
it accounts for only 8% of the animal's body mass.
Its spine is very flexible.
Its hind legs are ideal for jumping,
with short thighs and
the bones of the lower leg (fibula and tibia)
combined into one strong bone.

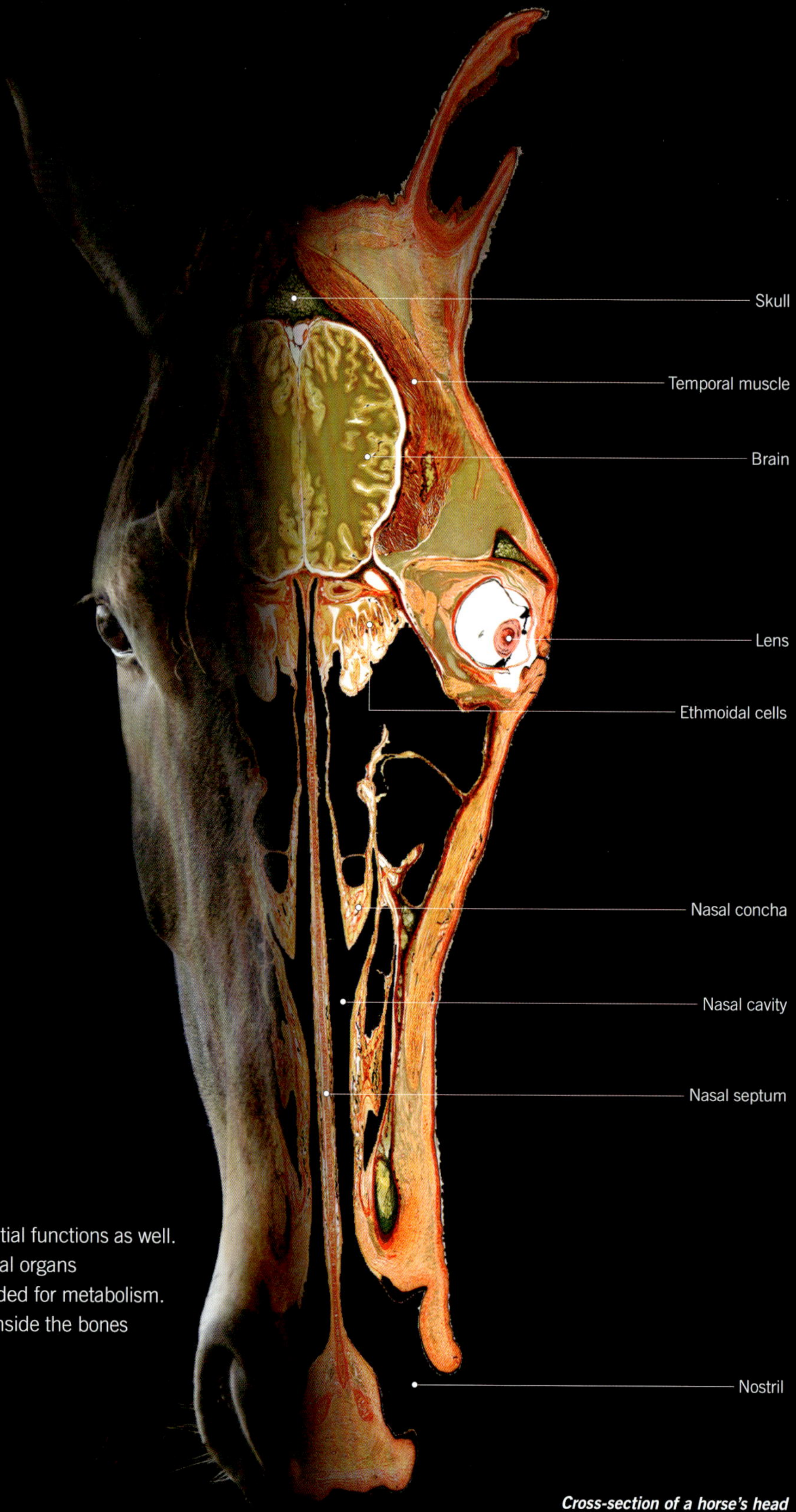

In many animals,
the bones have other essential functions as well.
They protect delicate internal organs
and store mineral salts needed for metabolism.
The bone marrow located inside the bones
produces the blood cells.

Cross-section of a horse's head
(Equus ferus caballus)

A vertebrate's skull contains and protects
the most important sense organs:
the eyes, nose, tongue, and,
at the base of the cranium, the inner ears
and equilibrium organs.
In addition, the skull determines the shape of the head
and houses the brain in the skull cavity.

Skulls usually consist of many joined individual bones.
In horses, the skull is made up of 37 bones.

In young animals,
these bones are connected by connective tissue,
allowing the skull and the brain to continue to grow.
Only when the animal has reached its full size,
the bones are joined.
The connective sites continue to be visible
as bone sutures.

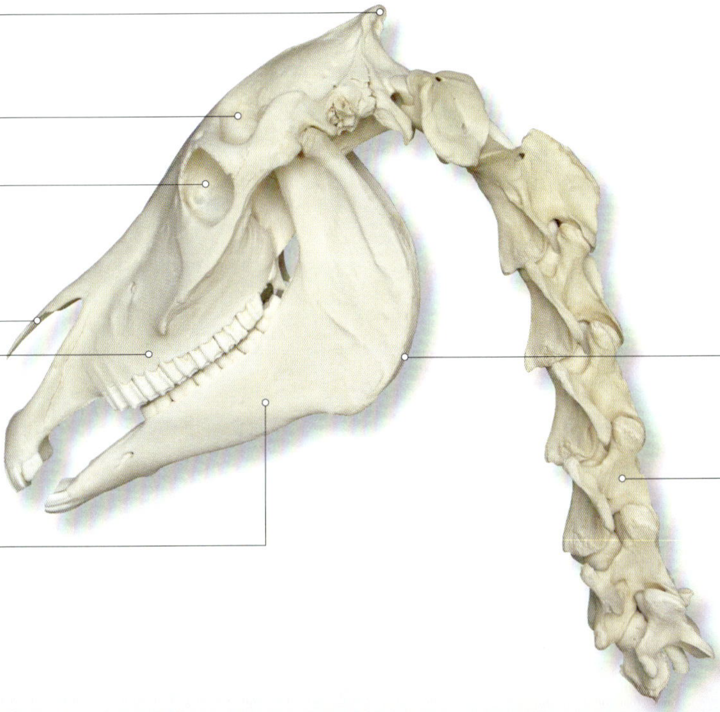

A horse's skull
(Equus ferus caballus)

This section shows the bones
of both the fore and hind leg in full length.
At their ends, they contain an area of hyaline cartilage,
the epiphyseal plate.
These plates permit growth after birth,
allowing the long bones of the body to extend
and the animal to grow as it reaches maturity.
In adults, these cartilage areas are replaced
by a fine line, the epiphyseal line.

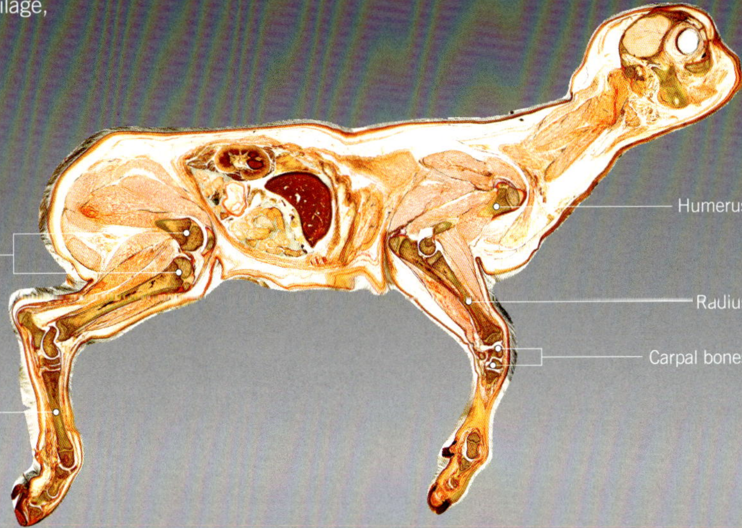

Longitudinal section of a lamb
(Ovis aries)

Skeleton of a mackerel
(Scomber scombrus)

Fish skeletons are made
either from cartilage (in cartilaginous fish)
or from bone (in bony fish).
The ribs are attached to the spine,
reaching around the entire body.
There is no breastbone.

Bony fish also have bony connective tissues
located inside the muscle sheaths, the so-called fishbone.
Fishbone structures are not attached to the spine.
They support the muscles of the body
without limiting their flexibility.

A skeleton may also consist of a stable shell,
such as in insects and crustaceans.
In these cases, it is called an exoskeleton.

The spider crab
has a strongly mineralised armour
that serves as an external skeleton (exoskeleton).

Burgundy snail
(Helix pomatia)

Snails have a shell that supports their invertebrate bodies.
They use many muscle contractions
to move smoothly and continuously across surfaces.
If it never stopped, a snail would take more than a week
to cover one kilometre (0.6 miles).
A suction effect
that is produced by sticky slime the animals secrete
allows them to hold on to different types of surfaces
and to even crawl overhead.

Scallops belong to the family Pectinidae.
Their outer housing consists of two calciferous shell halves
which are held together by the strong adductor muscle.

Scallop
(Pectinidae)

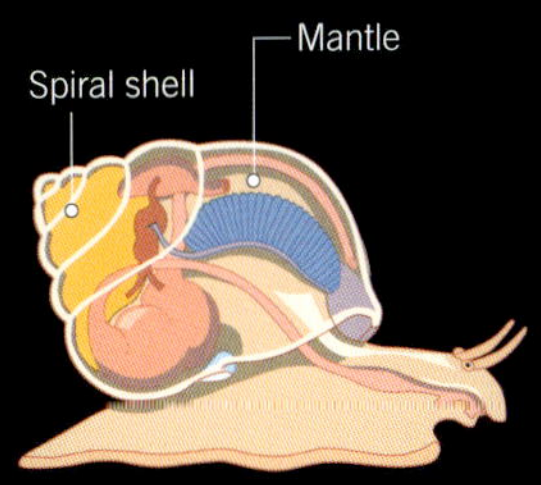

SNAIL

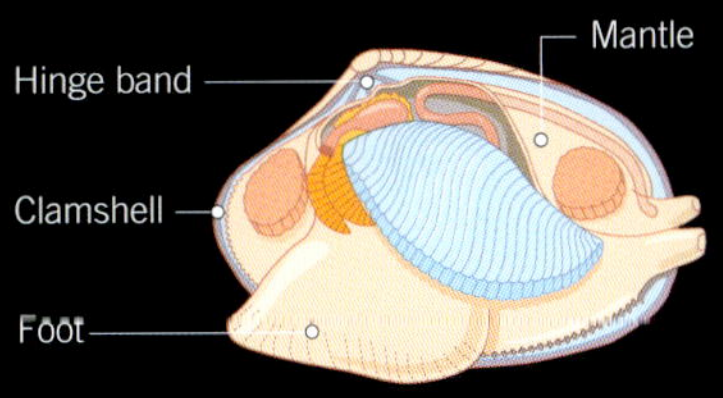

MUSSEL

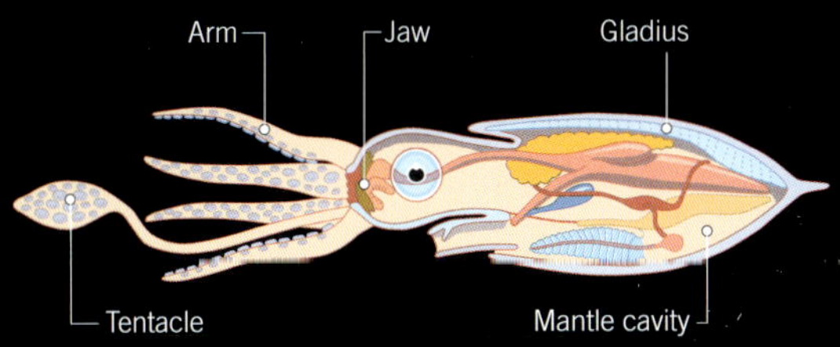

SQUID

A squid's body shape
is maintained by a flat horn endoskeleton,
the so-called gladius or pen.

*Thanks to its strong muscles,
a cheetah can achieve speeds
of 100 to 120 kilometres per hour (62–75 mph).
It is the fastest of all terrestrial mammals.*

Mice Under the Skin

In ancient Rome,
the visible motion underneath our skin when we move
was thought to be caused by tiny mice, *musculi* in Latin.
The English word *muscle* is derived from this early term.

Almost all animals have muscles.
They are the main drivers of movement
and allow animals to react to their environment.

Muscles fulfil their functions by contracting.
In doing so, they set parts of the skeleton
or other structures of the body in motion.

Mammals have more than 600 skeletal muscles on average;
some insects have three times as many.

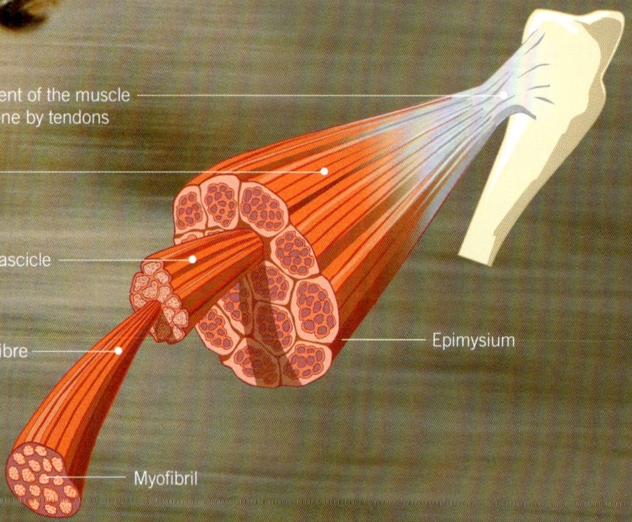

Various structures in the muscle
are bundled and packed in hard sheaths, like a thick cable.
The fundamental building block of every skeletal muscle
is the muscle fibre.
It consists of a cell that can be up to 15 cm (6 in) long
and about 0.1 mm (0.004 in) thick.
Inside there are threadlike myofibrils
that allow the cell to contract.
Several muscle fibres are combined into bundles
called *muscle fascicles*, in turn,
these are grouped together to form the whole muscle.
On the outside,
the muscle is surrounded by a thin layer of connective tissue,
called the *epimysium*.
Muscle fascia, a solid sheath of connective tissue
that keeps the muscle in its external form, is attached there.
The fascia continues as tendon at the end of the muscle belly.

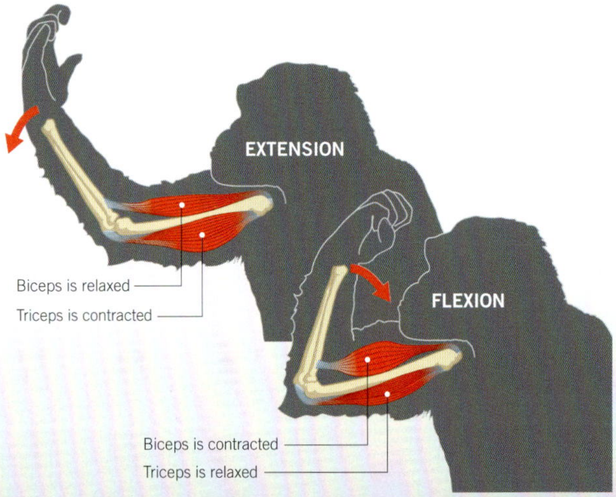

The biceps and triceps, two muscles located in the front extremities of vertebrates, demonstrate a simple antagonistic system, in which each muscle counters the pulling force of the other. If both muscles pull, the joint can be adjusted to and held in any position.

Muscles can pull in only one direction.
Therefore, most body movements
are not produced by a single muscle,
but rather by an entire group of muscles.
The individual muscles in these groups
often pull in opposite directions.

For example,
the biceps of the front extremities flexes the elbow joint,
while the triceps, acting as its antagonist, extends it.
Both work in perfect harmony – as one muscle shortens,
the other one counters the resulting movement,
preventing excessive deflection.
This enables flowing and controlled movement.

Specialised muscles also enable motion inside the body,
such as the heartbeat or the passage of food
through the gastrointestinal tract.
Its fibres run in the walls of hollow organs
and are controlled involuntarily.
That is, they are not controlled by the conscious mind.

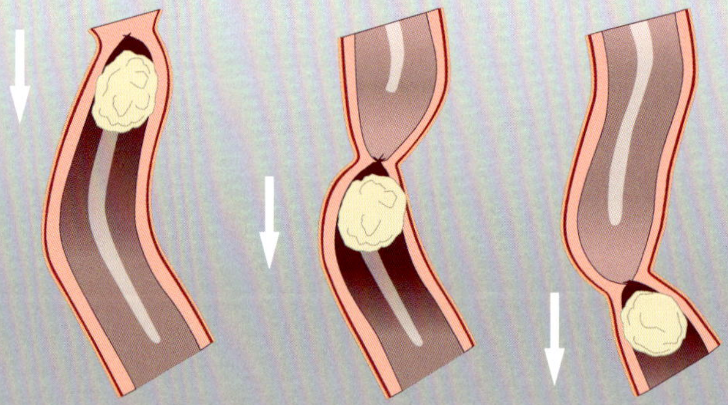

Food in the gastrointestinal tract is moved along by rhythmic, wave-like muscle contractions. This process is called peristalsis.

Longitudinal section of a duck
(Anas platyrhynchos domestica)

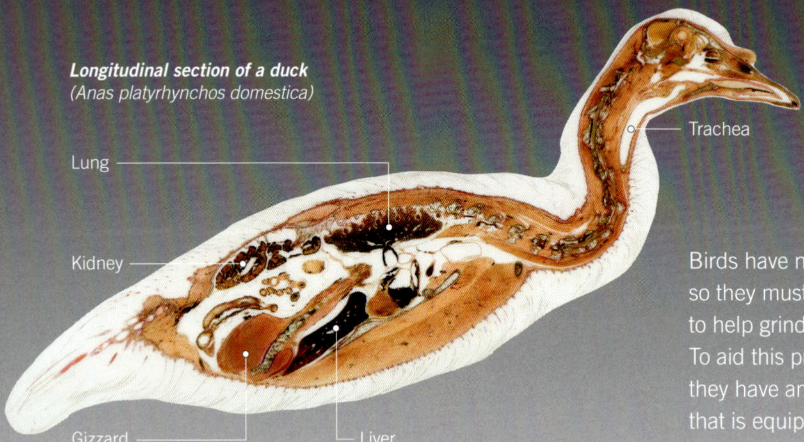

Birds have no teeth to chew with
so they must swallow small stones
to help grind food in the stomach.
To aid this process,
they have an additional stomach
that is equipped with a particularly strong muscle layer.

Ostriches are fast runners.
They reach speeds of up to 80 km/h (50 mph)
and are able to maintain them for about half an hour.
Very strong muscles in the back and legs facilitate this.
Long elastic ligaments act like springs.
They allow ostriches to regain energy
every time their feet push off the ground.
Unlike all other birds, ostriches have only two toes –
an adaptation to their high-speed running.

Ostrich
(Struthio camelus)

Electrified

Most animals have a fine network of nerve fibres
that extend into all regions of the body and
monitor and control its essential body functions.
The nervous system affects the entire animal,
controlling hundreds of activities simultaneously
including breathing, digestion,
movement coordination,
and the processing of all sensory perceptions,
such as seeing, smelling, tasting and hearing.
The nervous system also determines
how an animal responds to its environmental conditions.

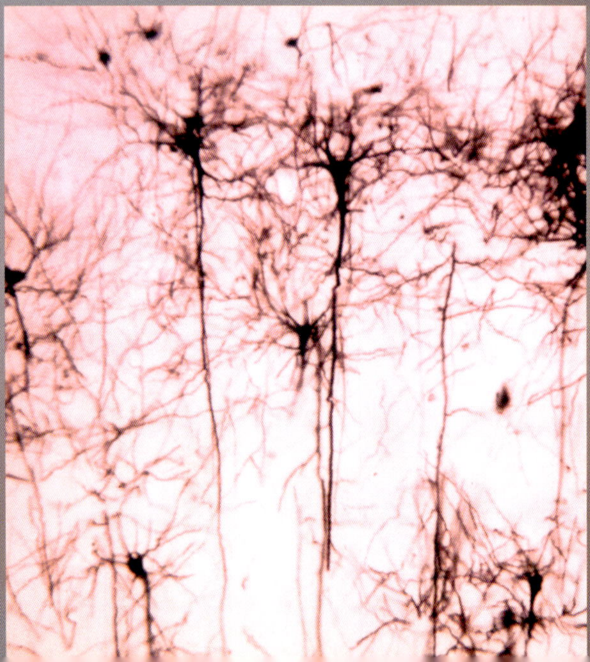

*In more complex developed animals,
several billions of nerve cells
jointly fulfil many specialised tasks.*

Sensory perceptions, such as smell, taste or touch, are processed in special areas of the brain. Tactile hairs enable animals to feel their way and to hunt even under conditions of poor visibility.

The nervous system of vertebrates comprises the brain,
the spinal cord and the peripheral nerves
branching off from the brain or the spinal cord.
These nerves branch out throughout the entire body,
growing finer and finer,
until they are no longer visible to the naked eye.

The brain is the overriding command centre.
It processes information from the sensory organs,
coordinates most movements,
and gives an animal the ability to feel, to remember,
and to communicate.

In many higher order animals,
the brain surface exhibits *sulci* and *gyri*,
or grooves and folds,
that have developed over the course of evolution
with the growth of the cerebrum.

The cerebellum and the brain stem
are located beneath the cerebrum.
The cerebellum is responsible for coordinating movement,
while the brain stem regulates many basic functions,
such as breathing and controlling the cardiovascular system.
The spinal cord emanates from the base of the brainstem.

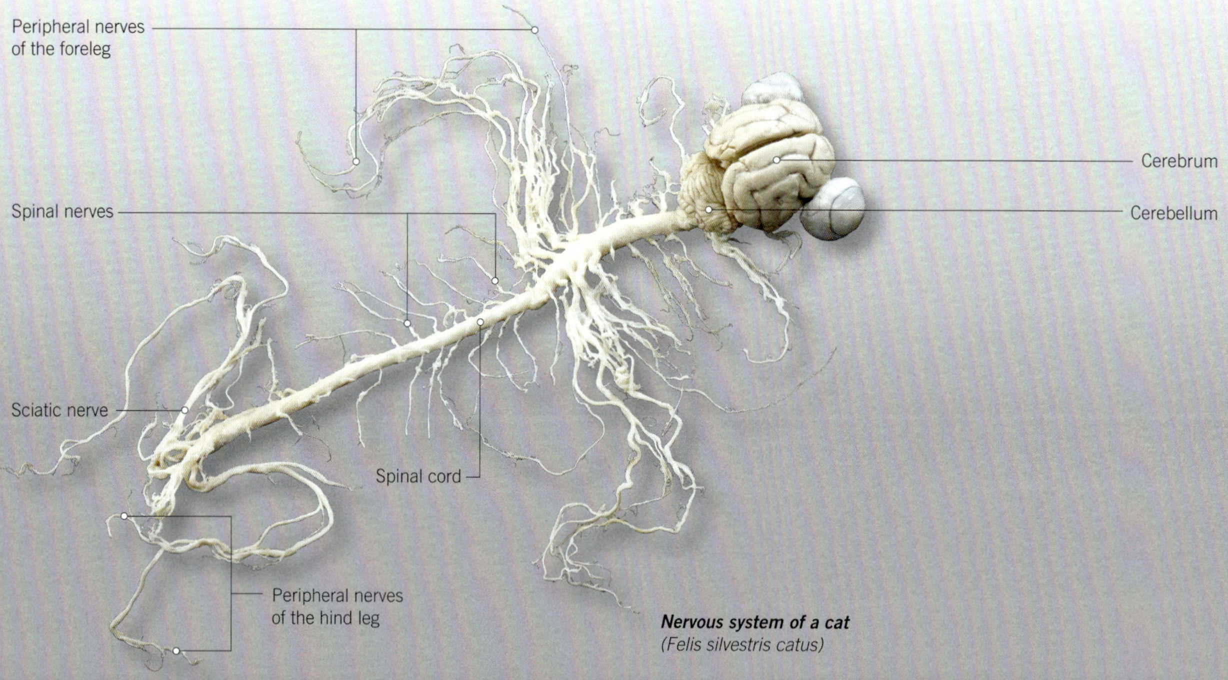

Nervous system of a cat
(Felis silvestris catus)

Monkeys exhibit the greatest intellectual abilities
in social behaviour.
The reason is probably their complex social life.
Monkeys must be able to figure out
the ever-changing social fabric of their group
in order to manipulate it if necessary.

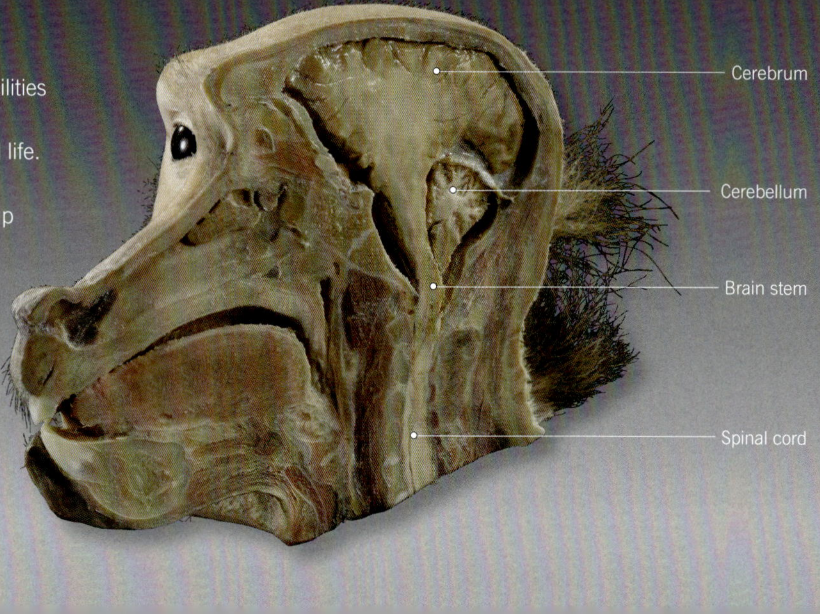

Half head of a baboon
(Papio)

Nerve cells, or neurons,
are the basic units of the nervous system.
The various regions of the body
communicate with each other
by means of weak electrical signals
emitted by neurons.
These signals travel at speeds of
up to 400 km/h (250 mph).

A typical neuron has a round cell body
with short extensions called dendrites.
The dendrites receive electrical signals from other neurons
and forward them via a longer extension.
This extension is called the axon
which can be up to 1 m (3 ft) long.
As a result,
electrical signals are transmitted over long distances.
The electrical signals are generated by the movement of ions
across the cell membrane.
Many axons have a fatty insulation *(myelin sheath)*
that is made up of so-called Schwann cells.

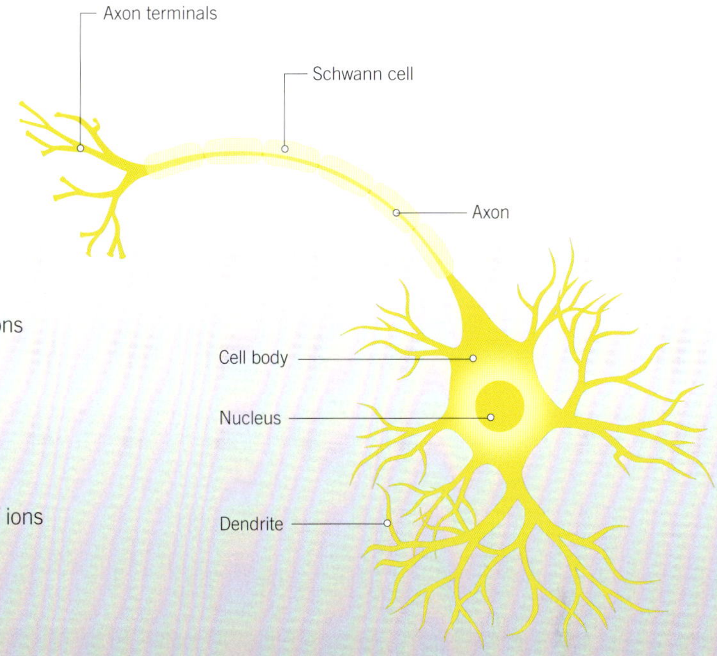

Schematic illustration of a neuron

Nervous Systems of Invertebrates

In invertebrates,
very different structures of the nervous system have evolved.
For example, hydra, freshwater organisms belonging to the
phylum Cnidaria,
have only a cobweb-like network of nerve fibres.
Arthropods, such as flies and other insects, have a brain
formed by partly fused ganglia and a central ventral nerve cord.

In molluscs, such as the snail,
the nervous system consists of different ganglia
that are connected by thick nerve fibre bundles.
Their neuron cell bodies
are linked by many short dendrites and axons.
This makes for an efficient exchange
and processing of information.

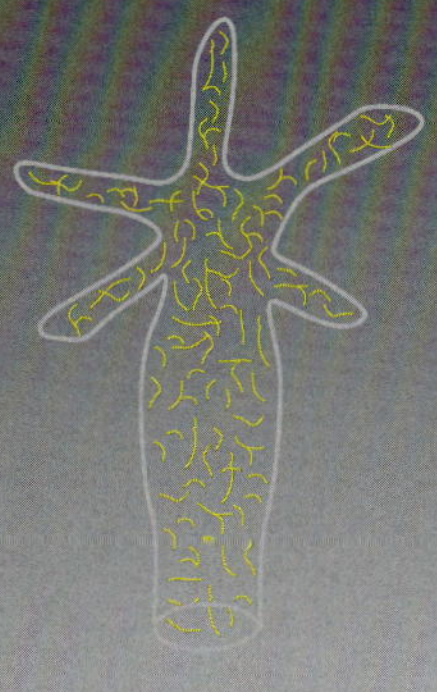

HYDRA

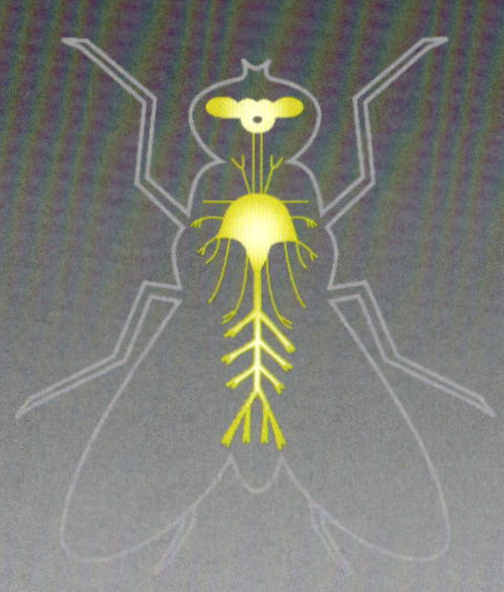

HOUSE FLY

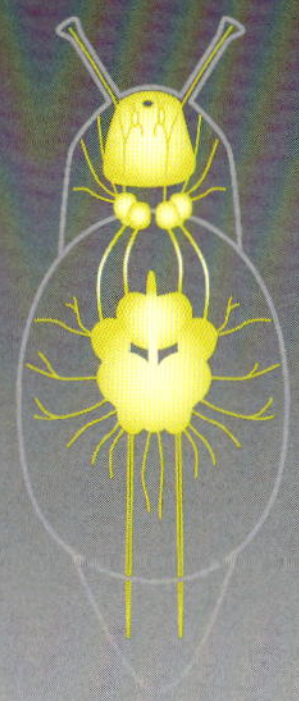

BURGUNDY SNAIL

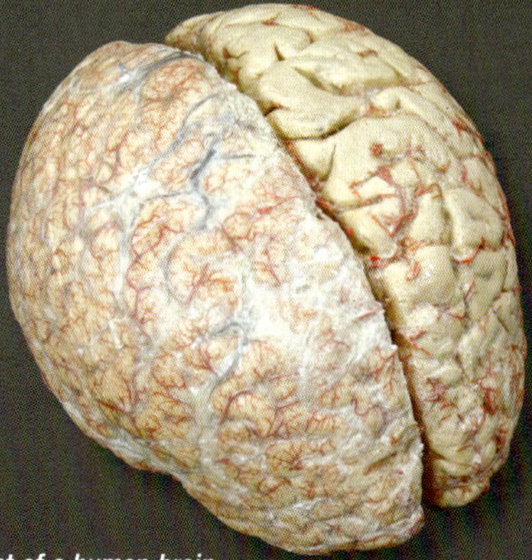

Average weight of a human brain:
1,300 grams (3 lbs)
(Homo sapiens sapiens)

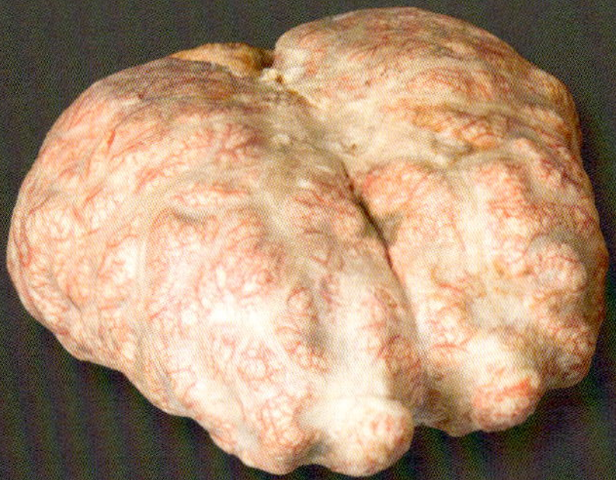

Average weight of a giraffe brain:
680 grams (1.5 lbs)
(Giraffa camelopardalis)

Average weight of an elephant brain:
6,000 grams (13 lbs)
(Elephas maximus indicus)

Usually, the larger an animal,
the larger the brain.
This is partly, because the brains of larger animals
simply have to control a larger body
and process more sensory information.
But that does not necessarily make them smarter
or more complex.
The size of the brain
does not offer any clues on intelligence

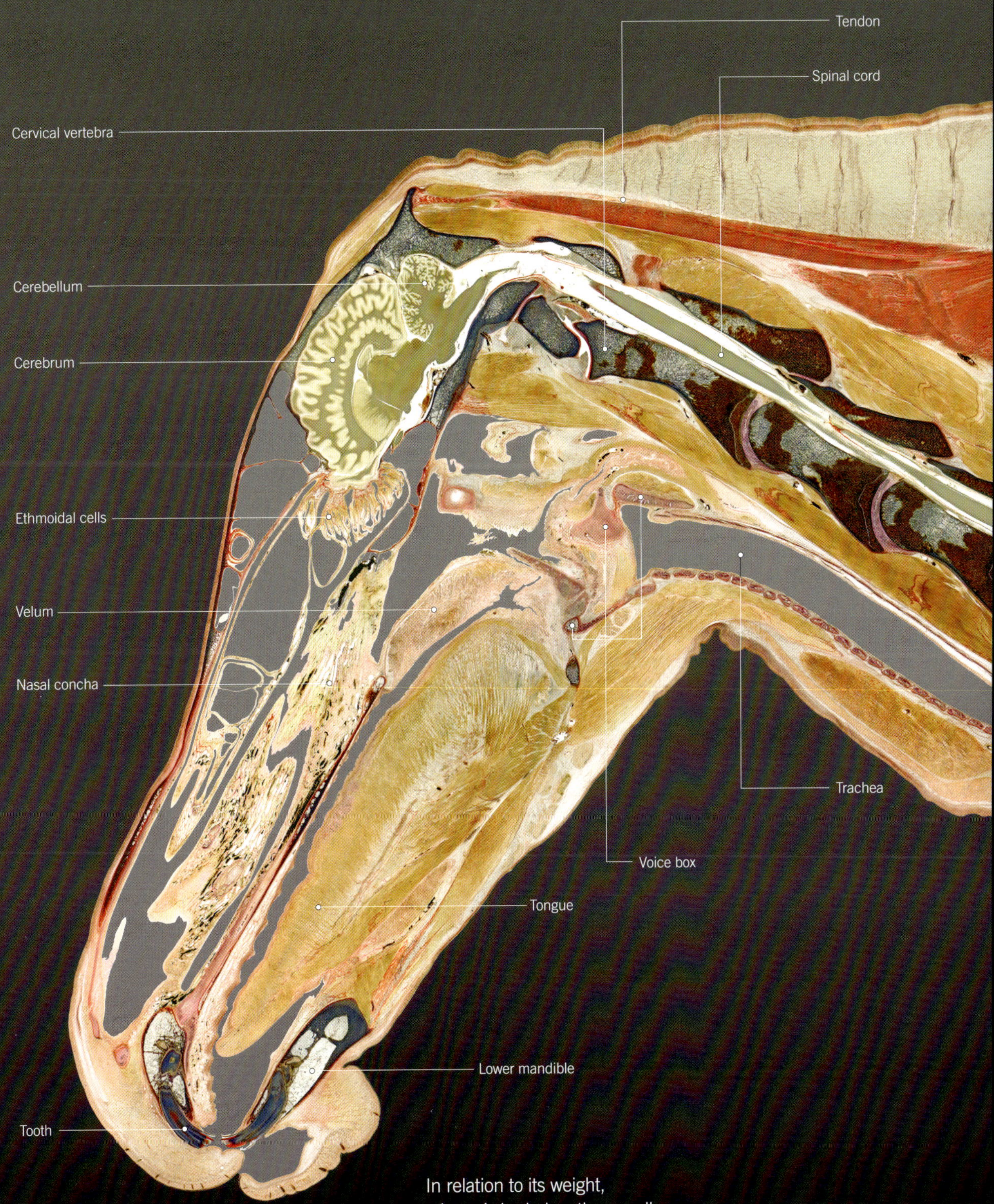

In relation to its weight,
a horse's brain is rather small.
It weighs between 400 and 700 grams (1–1.5 lbs).

Longitudinal section of a horse's head
(Equus ferus caballus)

Breath by Breath

Life requires a permanent supply of oxygen. Without this most cells of the body could not survive for more than a few minutes. Oxygen is indispensable for cell metabolism, a process that transforms nutrients into energy to keep the body functioning.

Most seals are capable of holding their breath for several minutes, some even for more than an hour.

Therefore, almost all vertebrates have lungs.
The lungs take up oxygen from the air
and pass it on to the blood.
At the same time,
carbon dioxide is released from the blood
and exhaled through the lungs.

Fish and many other animals living in water
have gills to enable this gas exchange.

Flatworm

Flatworms lack a circulatory system
to distribute the oxygen.
They do not have any special respiratory organs either,
such as gills.
Instead, flatworms obtain oxygen by diffusion.
The shape of their flat, leaf-like body
allows more oxygen to be absorbed through the skin.

The lungs of mammals constitute a large surface, where the transfer of oxygen into the blood takes place. First, air passes through the nose and mouth, then through the throat and trachea. The trachea branches out into two main bronchi, each leading into one lung. Similar to the branches of a tree, the bronchi taper into finer and finer branches.

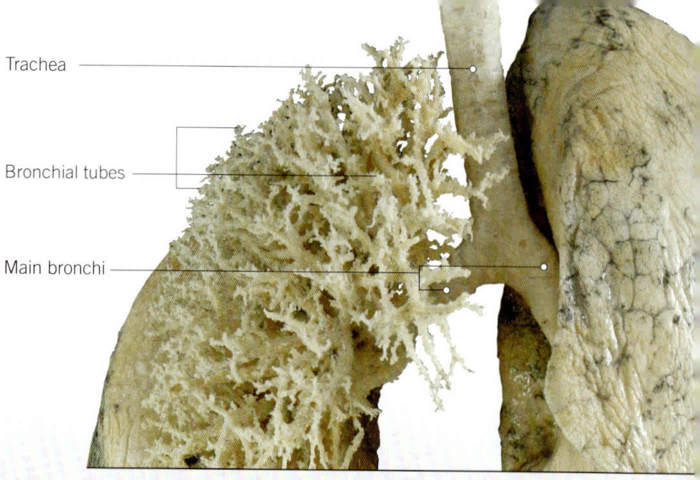

Plastinated human lung with bronchi
(Homo sapiens sapiens)

The smallest bronchi in the lungs end in tiny air sacs (alveoli) which give the lungs their sponge-like appearance. They are best recognised on an enlarged cross-section of the lung. A network of fine blood vessels (capillaries) surrounds the alveoli, where the gas exchange takes place. Oxygen from the inhaled air is released into the blood and carbon dioxide from the blood is released into the air that will be exhaled.

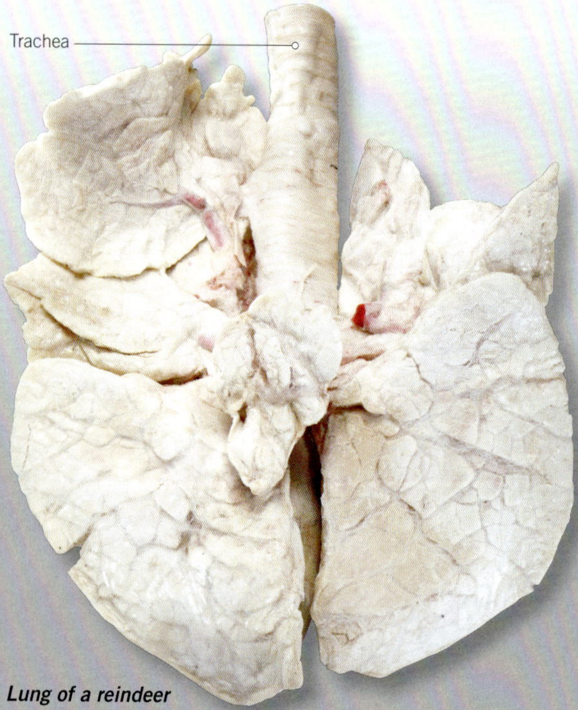

Lung of a reindeer
(Rangifer tarandus tarandus)

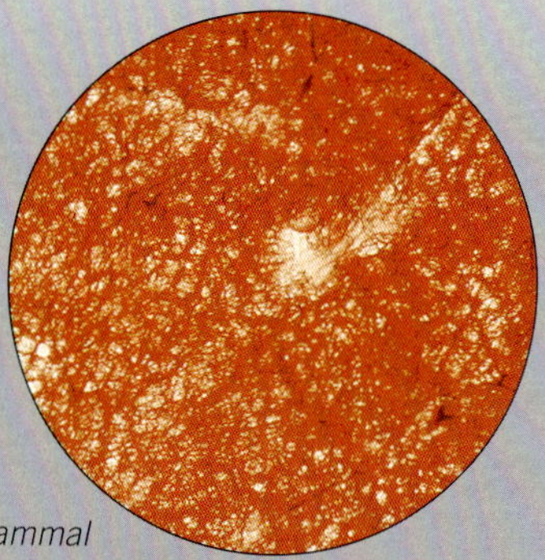

Lung tissue of a mammal

The lungs of birds are relatively small. Nevertheless, a bird can breathe in three times more air than a much larger mammal, because their respiratory system is an open tube system. Additional air bags act like bellows that constantly flush the lungs with air. The thin-walled air sacs take up considerable space in the body cavity of birds. They also extend into the large bones rendering the bird's skeleton lightweight.

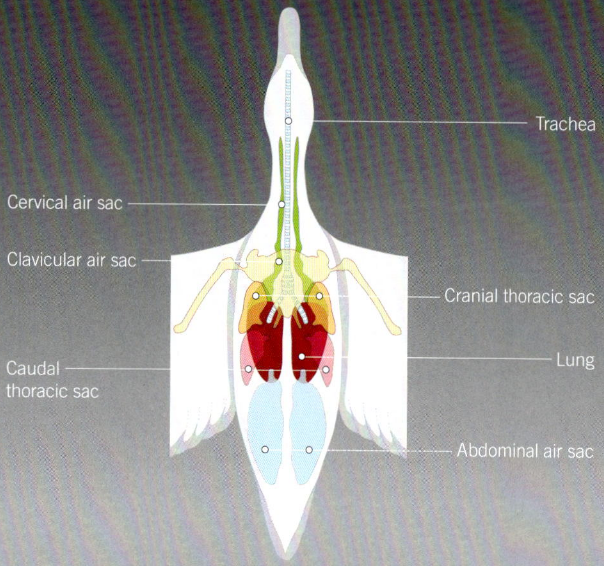

The lungs of mammals
are protected within the thoracic cavity.
When the chest expands, air flows into the lungs.
The main breathing muscle is the diaphragm,
a thin, dome-shaped muscular plate
stretching across the lower chest.
When its muscle fibres contract,
the diaphragm flattens out,
increasing the space in the chest
and the lungs themselves.
This causes pressure inside the lungs to fall
and air to flow in.

The external intercostal muscles support inhalation
by pulling the chest upwards and outwards.
The exhalation is hardly supported by the body.
The diaphragm and intercostal muscles
simply relax and thereby push out the air.

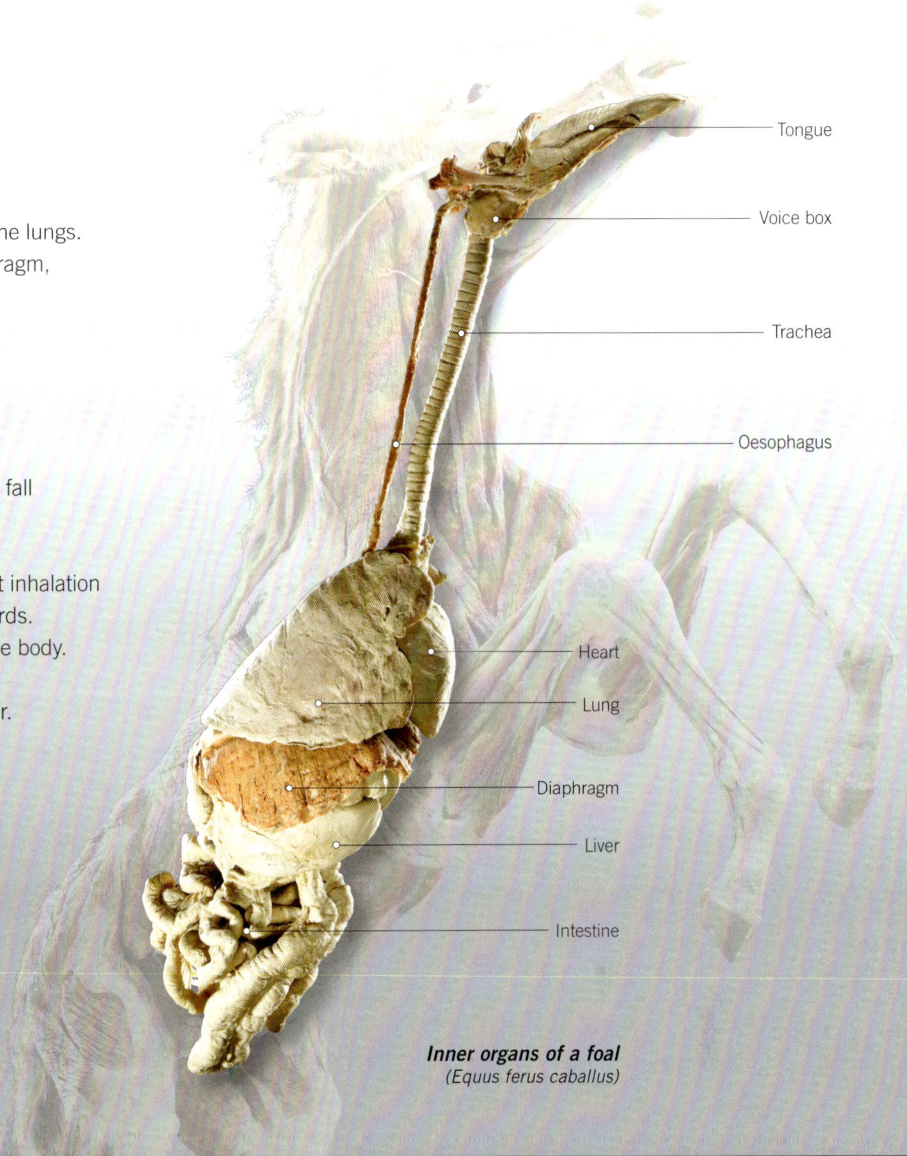

Inner organs of a foal
(Equus ferus caballus)

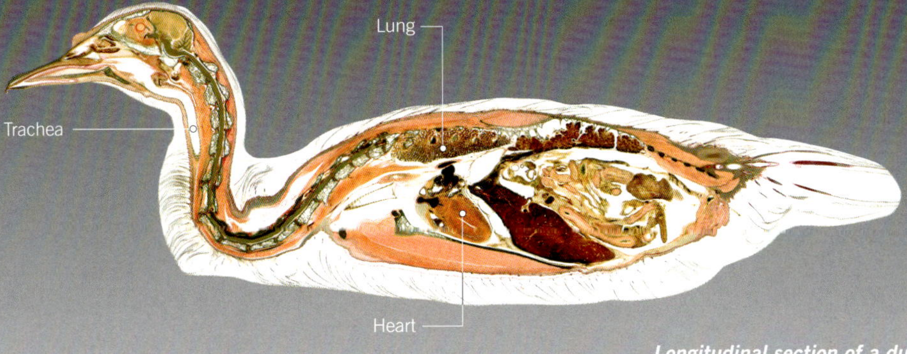

Longitudinal section of a duck
(Anas platyrhynchos domestica)

Gills are organs
that facilitate respiration in water.
Gills are also found in snails, mussels, clams
and other molluscs,
as well as in the larvae of amphibians.

Fish gills consist of
several bony and cartilaginous arches
located alongside the head.
They have large surfaces, amply supplied with blood,
that filter oxygen from the water
and excrete carbon dioxide.
For this exchange to work,
the gills have to be exposed to moving water.

In bony fish,
the gill chamber is enclosed by a lid (operculum)
that covers the gills.
It supports the intake of respiratory water
by a vacuum-suction system.

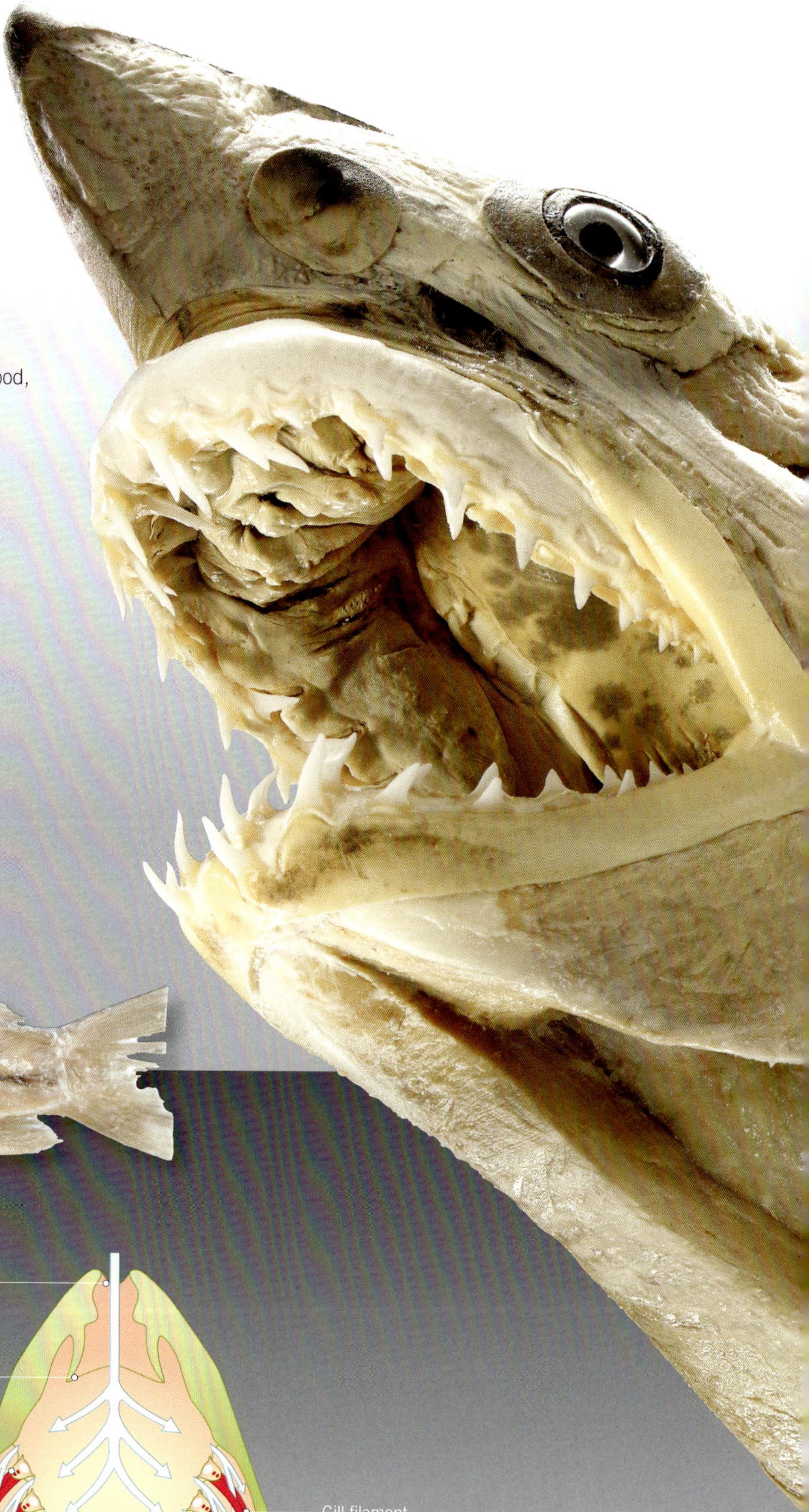

Mackerel with dissected gills
(Scomber scombrus)

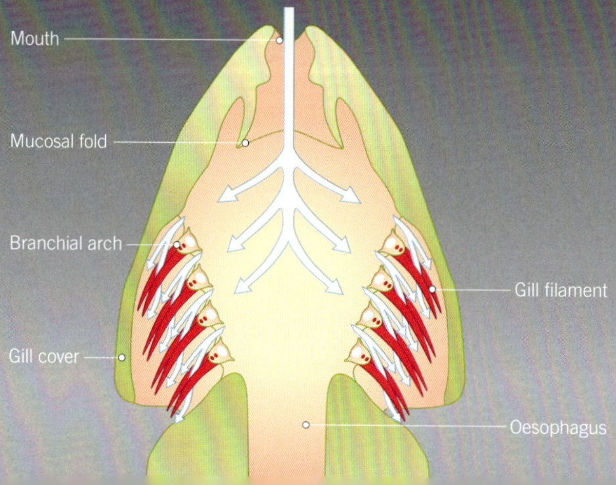

Cartilaginous fish, such as the shark, on the other hand, have no operculum. Their gill slits are open toward the outside. In order to still filter enough oxygen out of the water, the shark must create the necessary water flow through the gill leaves by his body movements. Therefore, most shark species can only breathe when they swim.

External Gills

Amphibian larvae, as this eft, carry external gills on the sides of the head. They are vulnerable but regenerate quickly.

Newt tadpole

Mako shark
(Isurus)

Network of Life

Large, complex animals
have an important internal transportation system:
the cardiovascular system.
It distributes nutrients, oxygen
and hormones to individual regions of the body,
and also collects metabolic byproducts,
which are then eliminated.

The heart is the engine of this system,
and the dense network of blood vessels
forms the transport routes.

The glass frog's heart is visible at the front of its chest.
The thick muscle walls hide the red blood inside,
but it is clearly visible in the abdominal vein leading up to the heart.

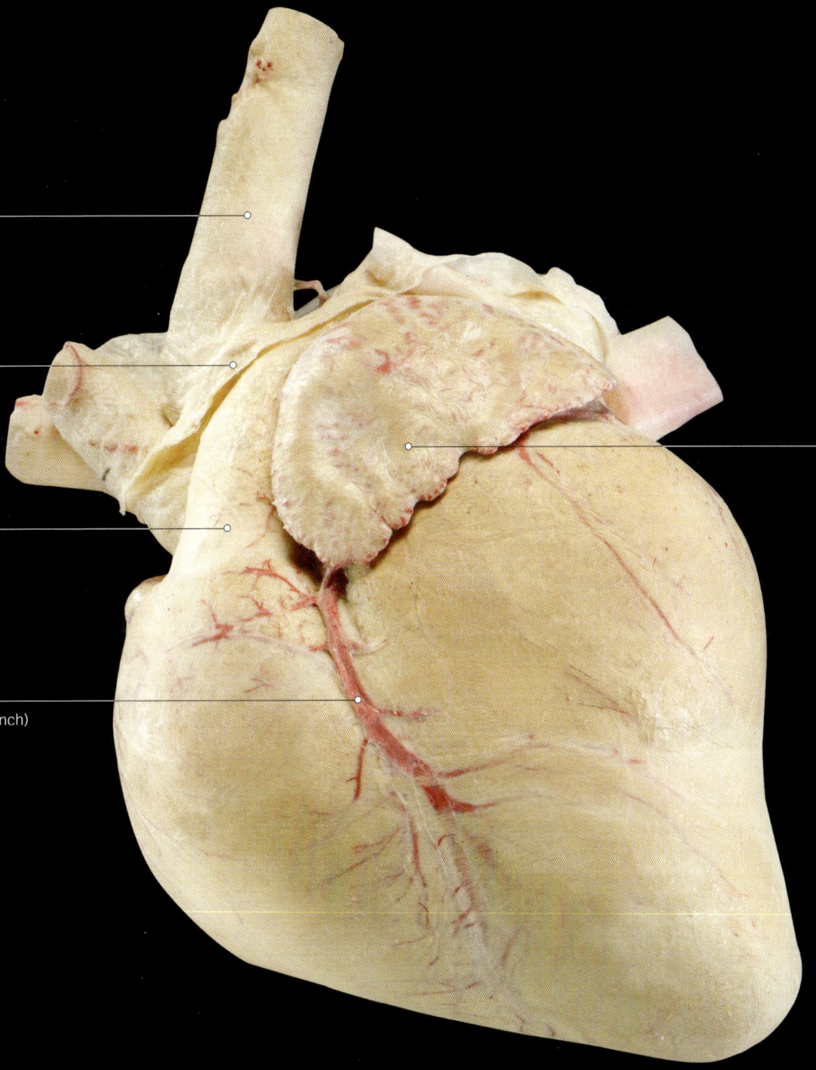

Heart of a reindeer
(Rangifer tarandus tarandus)

The heart is a hollow muscle organ
responsible for maintaining the circulation of the blood
through the body.
Its muscle fibres run in spirals.
When they contract,
the chambers of the heart are compressed,
pushing the blood into the vessels.

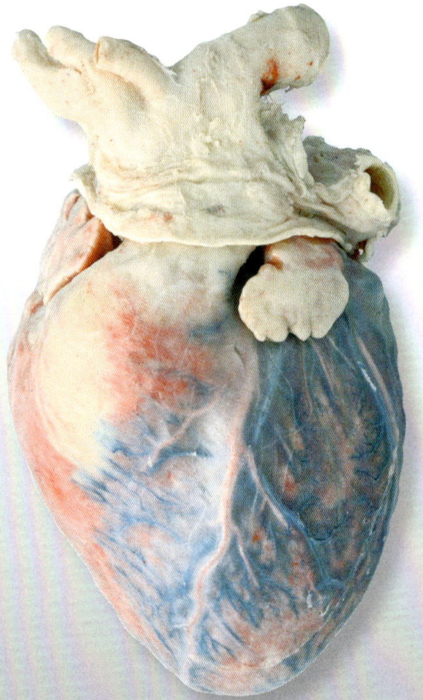

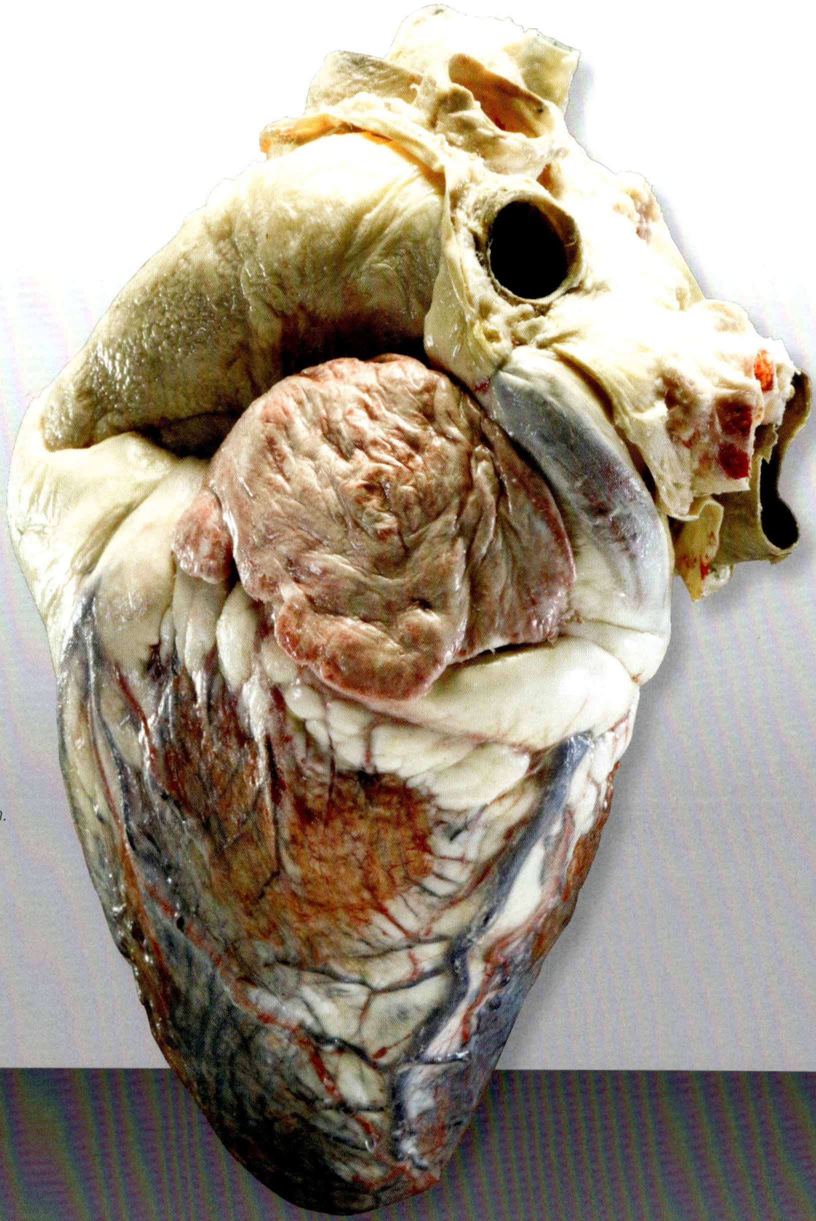

Human heart (left) and a cow's heart (right) in comparison.
In mammals, the size of the heart
correlates very much with the body weight.
The healthy heart of a person
makes up about 0.5% of the body weight
and weighs 300 to 350 grams (10–12 oz) on average.
A cow's heart weighs an average of 2.5 kilograms (7 lbs)
and makes up 0.4 to 0.6% of the body weight.

The hearts of mammals and birds exhibit a similar structure,
with four chambers and four valves.
The size of the heart
is roughly proportional to the weight of the body.
Thus, these hearts differ mostly in their size and weight.

Most reptiles, however, have a heart with two atria
and only a single main chamber without a septum.
The fish heart has the simplest structure
among vertebrates.
It only consists of a thin-walled atrium
and a thick-walled, muscular chamber.
Between the two, there is a flap
that prevents the backflow of blood.

Located in the upper part of the heart of mammals and birds, two atria capture the incoming blood. Below are the heart chambers (ventricles) whose muscle power is much stronger. The left ventricle pumps blood into the arteries of the large body cycle with relatively high pressure. The arteries lead into the capillary network of the organs. From there the blood is re-collected in the veins to be transported back to the right atrium. The right ventricle finally pumps the blood into the lungs for re-oxygenation (pulmonary circulation).

Since the total vascular resistance of the systemic circulation is significantly larger than that of the pulmonary circulation, the left ventricle has to build up significantly more pressure. Therefore its walls are much thicker than those of the right ventricle. However, the filling volume and cardiac output in both ventricles are the same.

The blood flow inside the heart is regulated by valves consisting of fibrous connective tissue. Because of their shape they are called atrioventricular valves. At the openings to the aorta and pulmonary artery, pocket-like flaps prevent the return of blood flow into the ventricles. The flaps clashing during cardiac action produce the typical heart sounds.

A mammalian heart will beat about 1 billion times during the animal's lifespan, independent of the animal's size and length of its lifespan. The shrew, for example, lives in the fast lane. Its heart beats about 1,000 times per minute. Thus, it uses all its available heartbeats within about 2 years. Bigger animals have a slower metabolism. They breathe more slowly and have a lower pulse. For instance, the heart of an elephant beats only about 30 times per minute. Elephants can live up to 70 years.

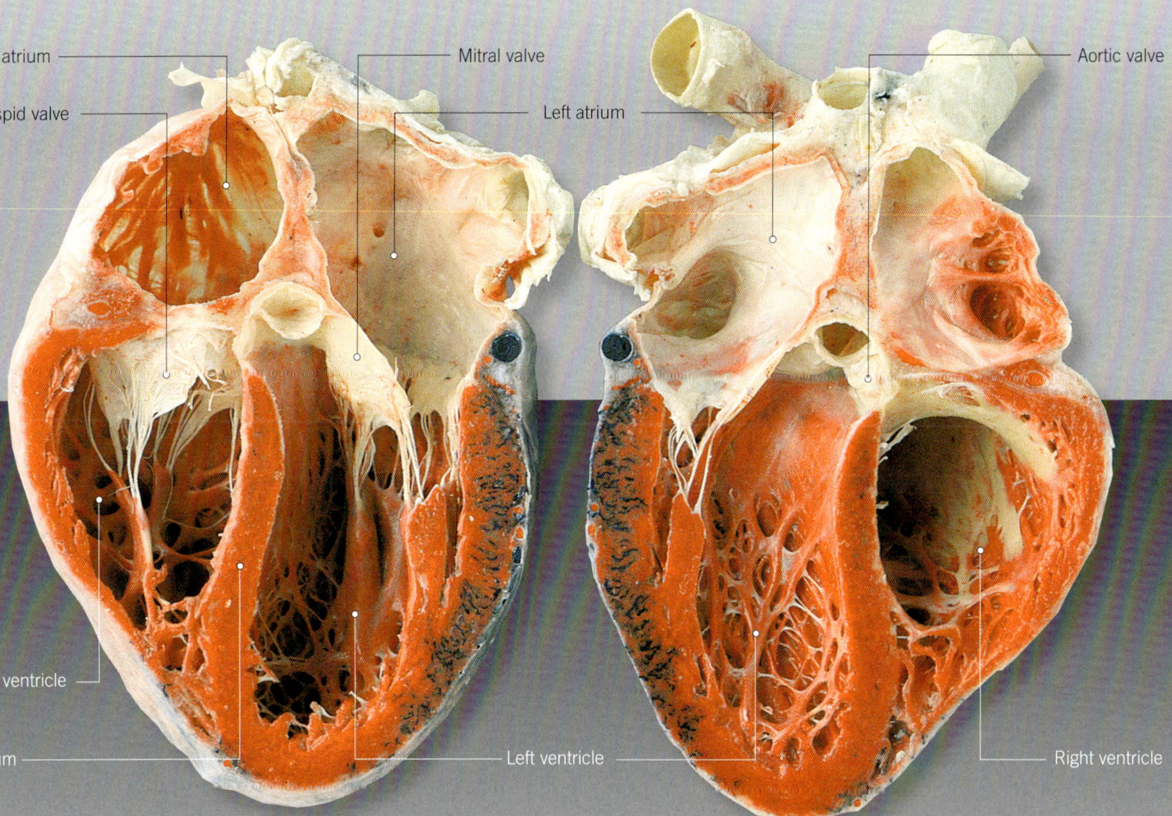

A human heart is about the size of a fist. At rest, it beats about 70 times per minute and pumps about 75 millilitres (2.5 oz) of blood with each beat. That amounts to about 200 million litres (53 million gal.) of blood pumped in an average lifetime of 75 years – enough to fill more than three supertankers.

Blood Vessel Configurations

Blood is carried from the heart through the entire body and back in three types of vessels:
the arteries, the veins and the capillaries.

Arteries are the vessels leading away from the heart.
They are like major highways
that distribute the blood across the body.
The aorta is the largest artery in the body.
It ascends directly from the heart.
The arteries branch off repeatedly,
getting smaller and smaller in diameter.
When they reach their target organs and tissues,
they are only as thin as hair and are called capillaries.
It is through the walls of the capillaries
that the interchange of nutrients, oxygen
and other substances takes place between blood and tissues.
Then the blood collects in veins
that coalesce into bigger and bigger vessels,
carrying the blood back to the heart.

Blood vessel configuration of a rabbit
(Oryctolagus cuniculus forma domestica)

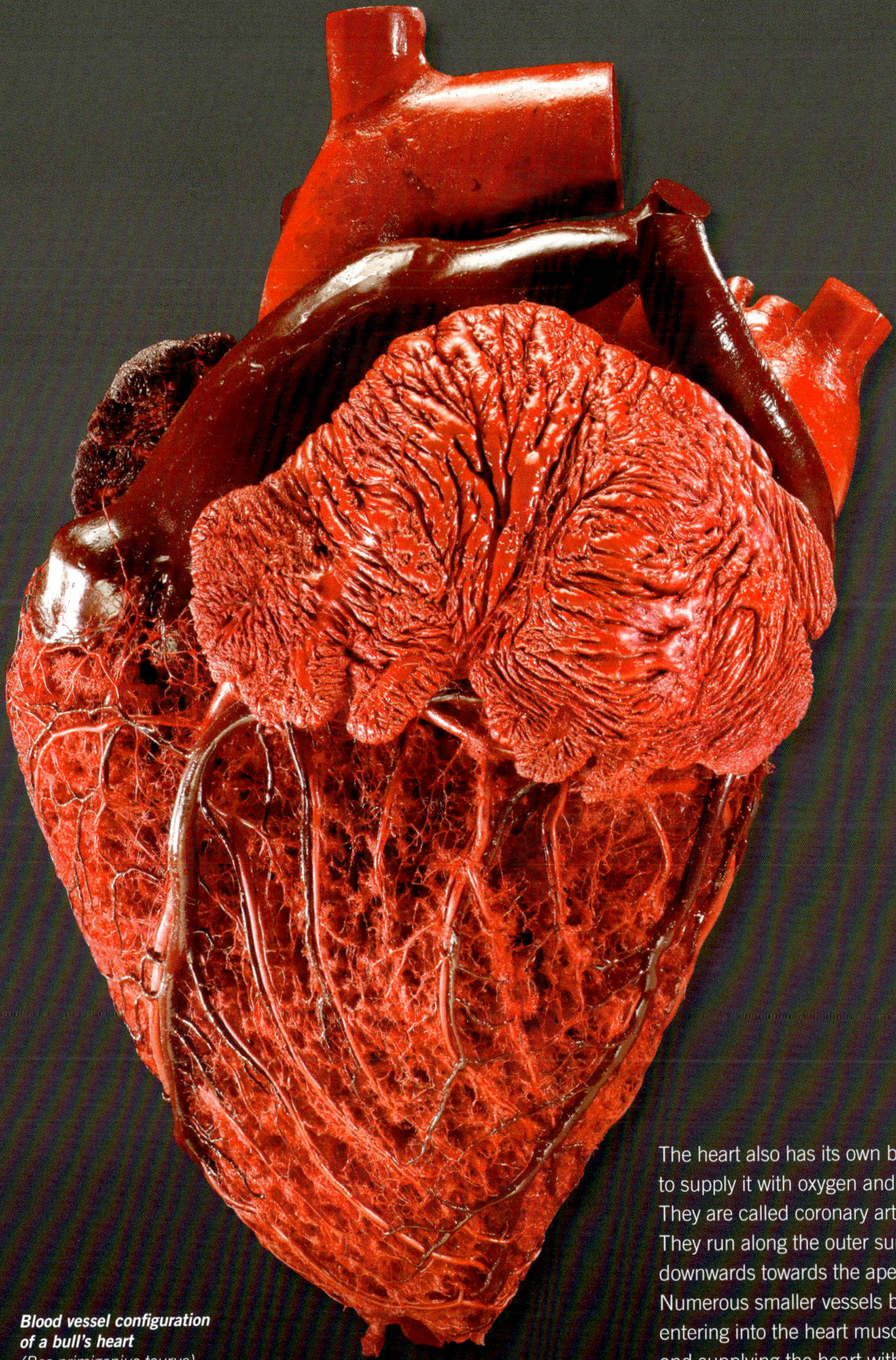

Blood vessel configuration of a bull's heart
(Bos primigenius taurus)

The heart also has its own blood vessels
to supply it with oxygen and nutrients.
They are called coronary arteries.
They run along the outer surface of the heart,
downwards towards the apex of the heart.
Numerous smaller vessels branch off them,
entering into the heart muscle
and supplying the heart with blood.
In this specimen,
both the vessels and the chambers of the heart
are shown as blood vessel configurations.

Blood vessel configurations are perfect representations of the internal cardiovascular profile.
They are created by injecting liquid coloured plastics into the vessels.
Once this plastic hardens, it assumes the shape of the vessels.
Then the surrounding tissue is removed both mechanically and with the help of enzymes.
Thus, blood vessel configurations are a cast of the vessels that have been injected with plastic polymers.

Only the main arterial branches are shown here in the images of the duck and the pig.
If the smallest vessels, the capillaries, were included as well, the vessel network would be too dense to see through.

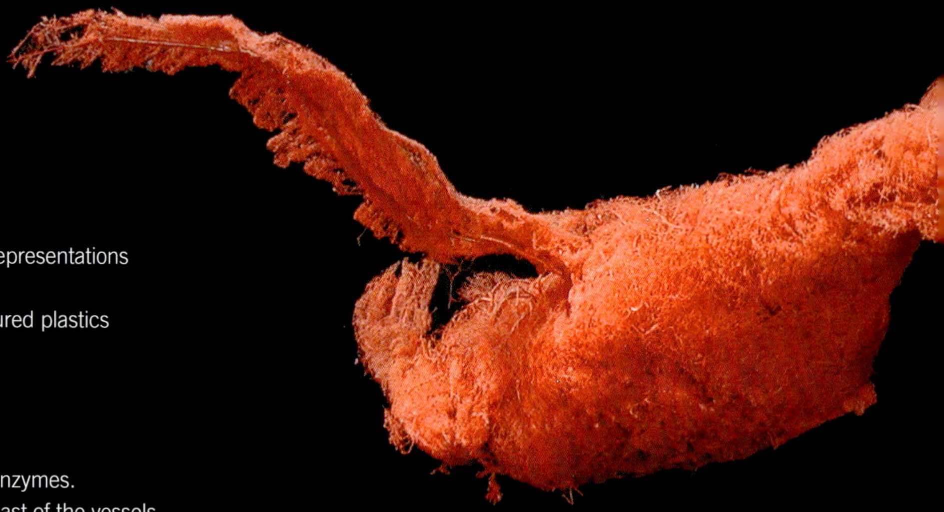

Blood vessel configuration of a flying duck
(Anas platyrhynchos domestica)

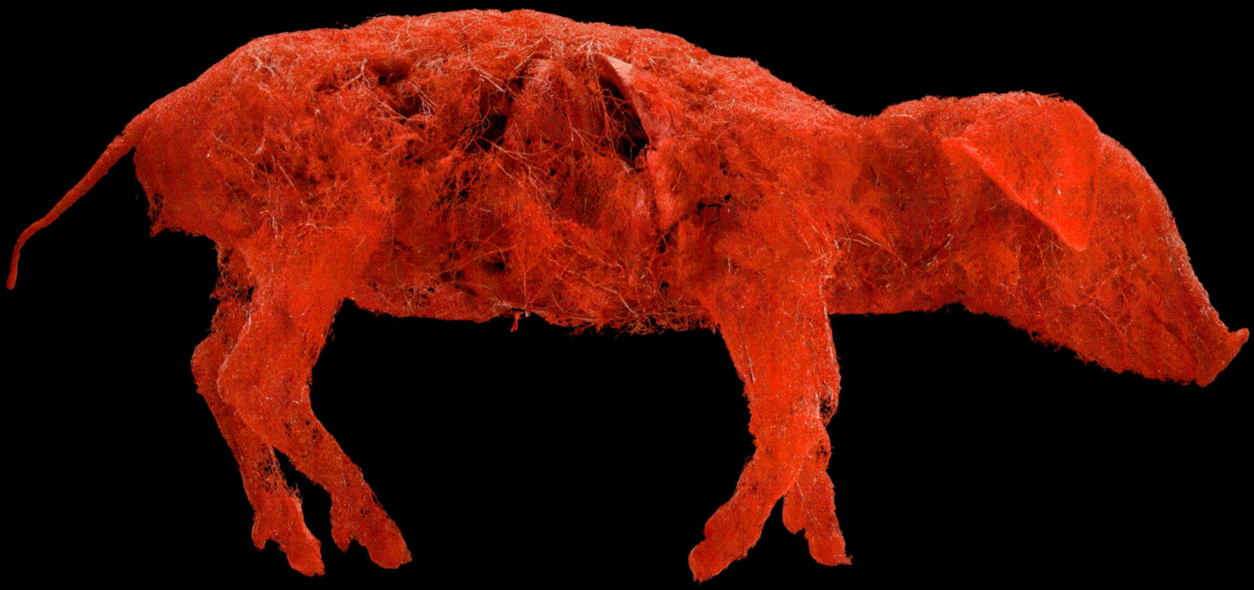

Blood vessel configuration of a pig
(Sus scrofa domestica)

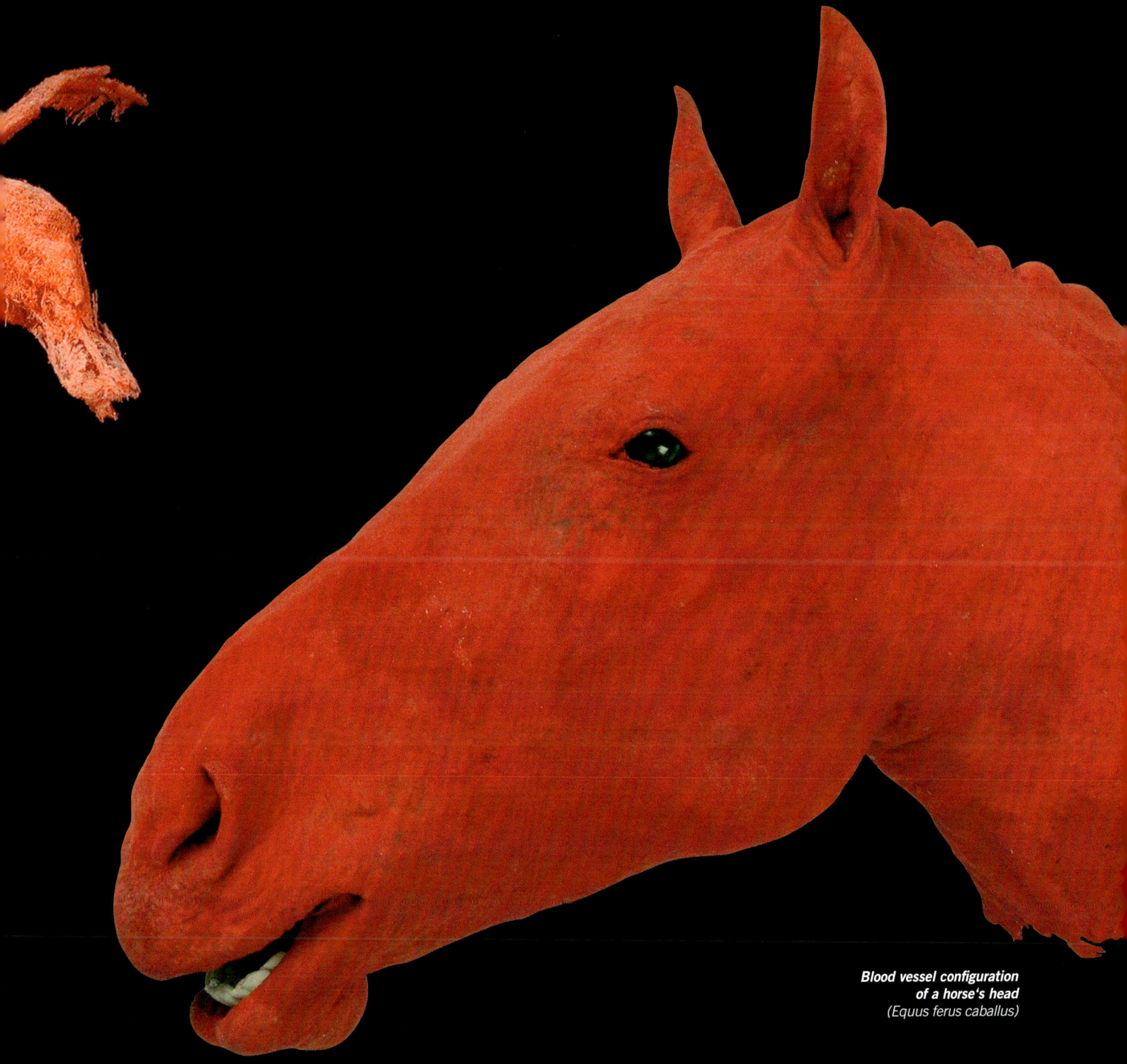

Blood vessel configuration of a horse's head
(Equus ferus caballus)

In this specimen,
the polymers have even penetrated the skin capillaries.
It shows how dense the vessel network of arteries
and capillaries actually is.
If you look at it with your naked eyes,
it seems as if you were looking at the horse's red-dyed skin.

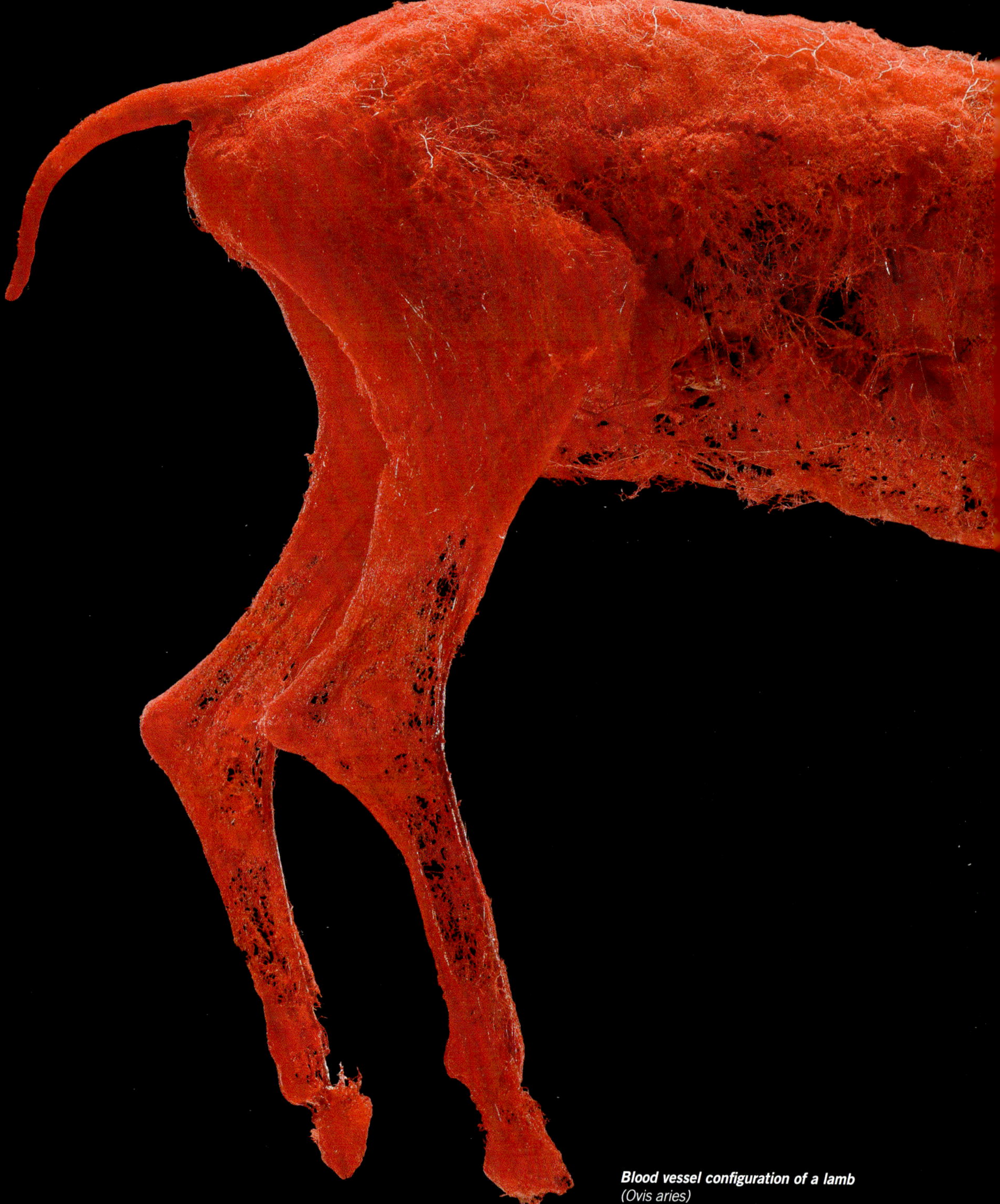

Blood vessel configuration of a lamb
(Ovis aries)

Blood – Elixir of Life

The blood of vertebrates has many functions:

- it distributes nutrients, hormones, and other messenger substances,
- it collects metabolic waste products,
- it serves to defend the body against diseases and
- it regulates the body's temperature.

Blood consists of a colourless fluid, the plasma,
which contains billions of red and white blood cells and platelets.
The red blood cells – also called *erythrocytes* –
are by far the most numerous.
They transport oxygen and carry away carbon dioxide.
The colourless white cells – the *leukocytes* –
are part of the body's built-in defence mechanisms.
The platelets, also called *thrombocytes*,
are involved in blood clotting.

In most vertebrates,
the blood flows so quickly
that one red blood cell is able
to travel through the entire body
in only a few seconds,
reaching even the most remote locations.

Fuelling the Body

All processes that occur in an organism require energy which is derived from food.
More developed animals have a complex digestive system that breaks food down in particles small enough to enter the blood stream.

Simply put,
the digestive system consists of a tube system beginning with the mouth and ending with the anus.
Depending on the size of the animal,
the digestive system may be several metres long.
The longest digestive tract is found in the sperm whale, at a length of up to 650 metres (0.4 miles).
In humans it is about 7 to 9 metres (20–30 ft) long.

Indigestible materials are excreted by the organism

Many animals have adapted to a special diet.
This makes them dependent on the geographical distribution of their food, as well as on its availability.

The digestive system of animals
is perfectly adapted to their diet.

Most vertebrates begin the digestive process
by roughly breaking down their food in their mouths.
From there
the food travels through the oesophagus to the stomach,
where digestion continues.

From the stomach,
the partially digested food moves into the small intestine
where digestive juices from the liver and pancreas are added.
The small intestine is the main site of digestion.
It is here that most of the nutrient molecules
pass through the intestinal wall into the blood stream.

Indigestible materials pass into the large intestine
where water is extracted
before it's expelled through the rectum as faeces.

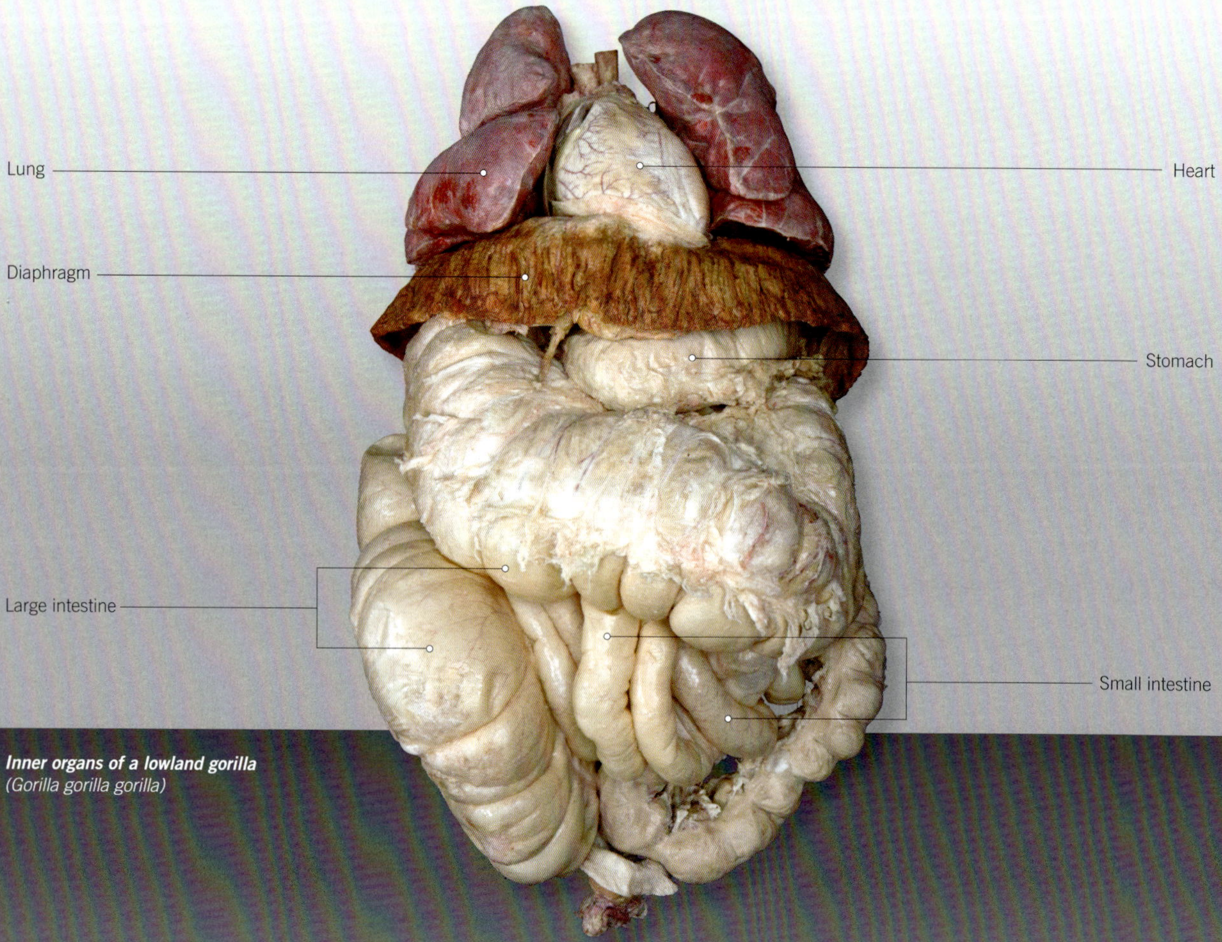

Inner organs of a lowland gorilla
(Gorilla gorilla gorilla)

This specimen
shows all the organs of the chest and abdominal cavities
in their natural position.
The chest and abdominal cavities are lined by a fine,
super smooth membrane, the *pleura* or *peritoneum*, respectively.
It also encases the organs almost completely.

Its secretions ensure
that the organs easily glide against each other.
The abdominal cavity houses the long intestines
typical of herbivores or plant eaters.
Gorillas spend half of their waking hours feeding.

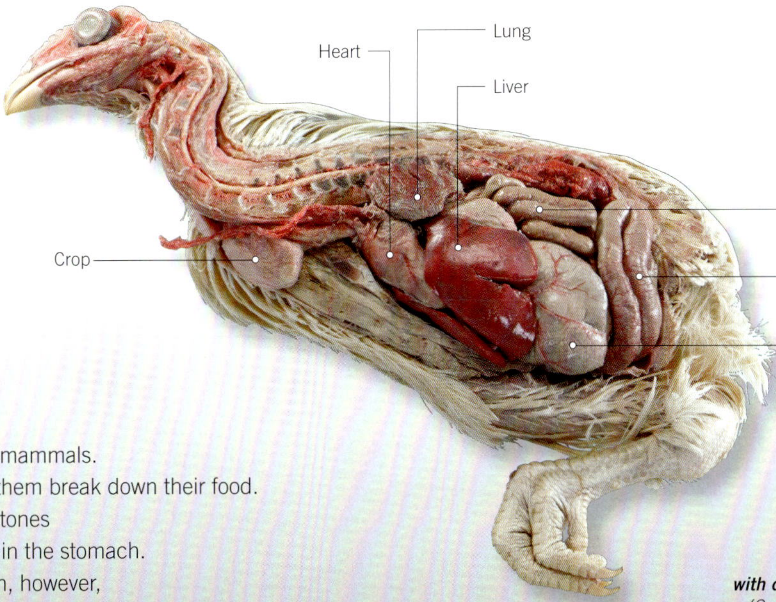

Chicken
with dissected inner organs
(Gallus gallus domesticus)

The digestive tract of birds is very different from that of mammals. Birds have no teeth to help them break down their food. Instead they swallow small stones that aid in grinding the food in the stomach. Before reaching the stomach, however, a bird's food is stored intermittently in the *crop*, a muscular pouch near the throat. Their intestines, in relation to the length of their bodies, are much shorter than those of mammals.

Digestive tract of a chicken

Anatomically less complex animals, for example jellyfish, sea anemones, or coral polyps, have only one body opening for the intake of food and excretion of waste.

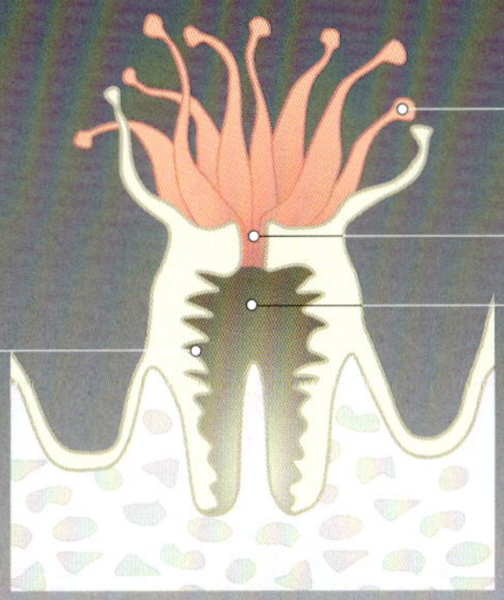

Coral polyp

Reindeer
(Rangifer tarandus tarandus)

Ruminants are mammals
that have a specially adapted digestive system
for the digestion of plant material.
Their stomachs have four chambers:
the rumen, reticulum, omasum and abomasum.
Food is initially broken down by microbes,
then regurgitated and chewed again
before it is actually digested.
This allows ruminants
to digest a wider range of carbohydrates, like cellulose
that are indigestible for other mammals.
Ruminants include not only cattle, sheep and goats,
but also deer, reindeer, antelopes, giraffes and gazelles.

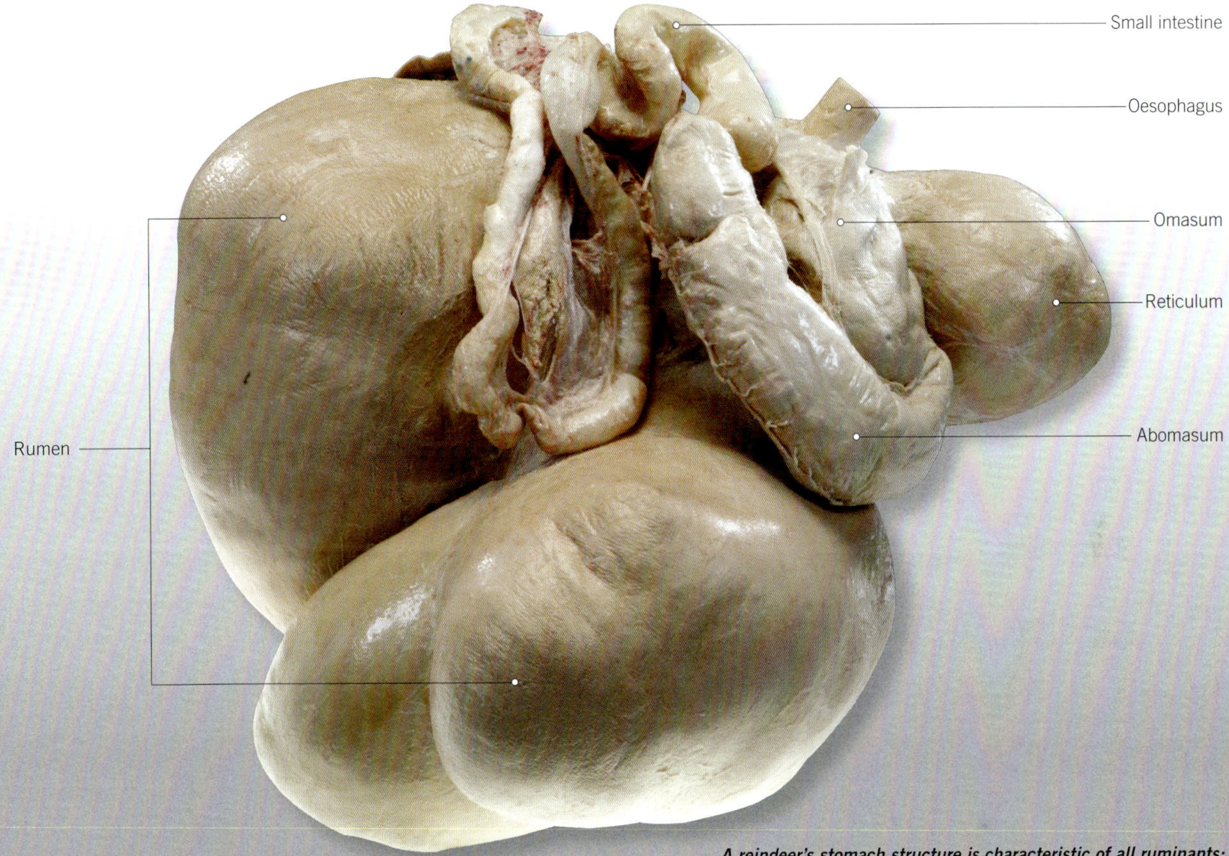

A reindeer's stomach structure is characteristic of all ruminants: it has four chambers consisting of the rumen, reticulum, omasum and abomasum.
(Rangifer tarandus tarandus)

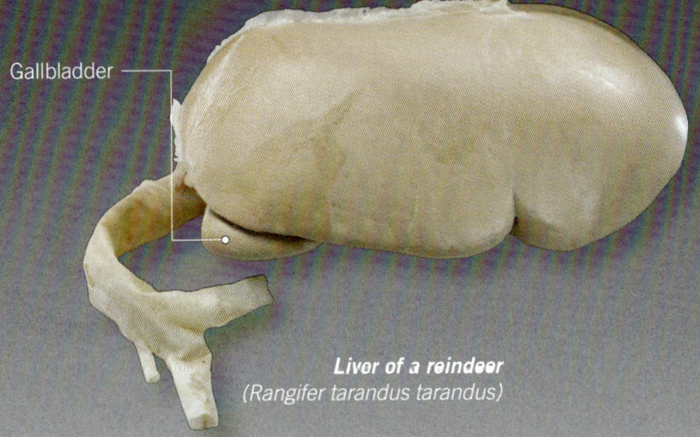

Liver of a reindeer
(Rangifer tarandus tarandus)

In vertebrates,
the liver is the central organ of metabolism.
Most importantly, it breaks down the nutrients
which it receives via the bloodstream from the intestine.
Hence, it assumes a key function
in the metabolism of carbohydrates, fats and proteins.
In addition, it acts as a water treatment plant
by absorbing toxins from the blood
and converting them to waste products.
The liver is the largest gland in vertebrates.

Ruminants mainly eat grass
which contains only few nutrients
and a lot of indigestible cellulose.

The food first reaches the rumen
that can contain up to 180 litres (47.5 gal).
It acts like a large fermentation chamber,
in which the food is broken down chemically
by microorganisms.
This process is called fermentation.

Then the food passes into the reticulum.
It plays an important role
in separating solid from already fermented food substances.
Solids are returned to the rumen for regurgitation.
Already fermented food substances, however,
are passed into the omasum.

In the omasum,
water and water-soluble nutrients are absorbed.
The thickened food then travels
into the actual glandular stomach, the abomasum.
There the food is mixed with hydrochloric acid
and the body's own enzymes for further digestion
and to enable its absorption of nutrients in the intestine.

The entire digestive process
takes ruminants two days.

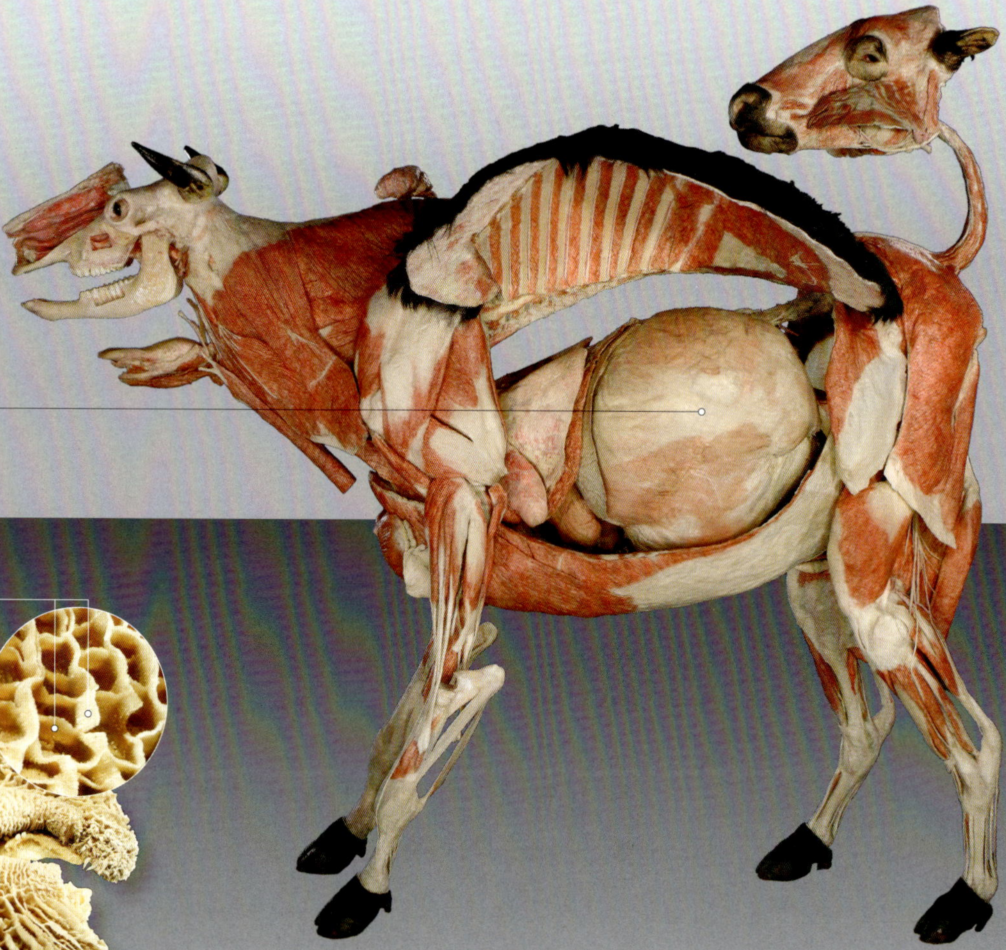

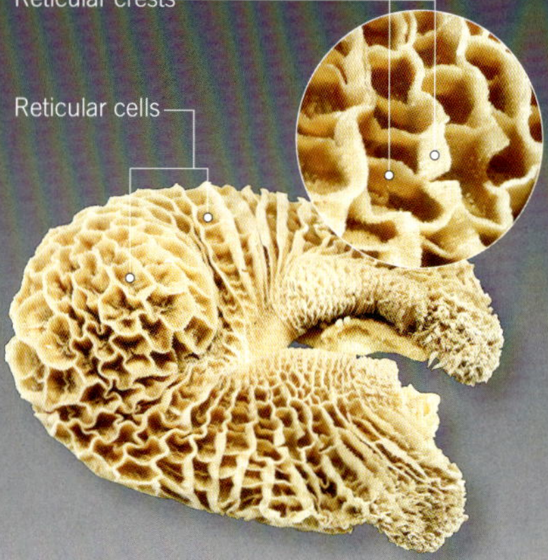

Interior view of a ruminant's stomach

A yak is a kind of ox that is endemic in Central Asia. Here the body cavities are opened widely to show the internal organs in their natural position. The large rumen takes up nearly the entire left abdomen.

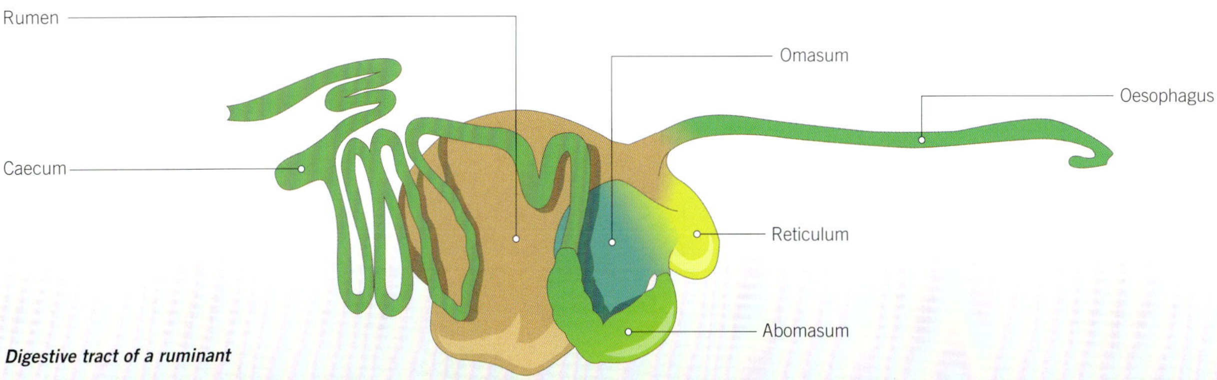

Digestive tract of a ruminant

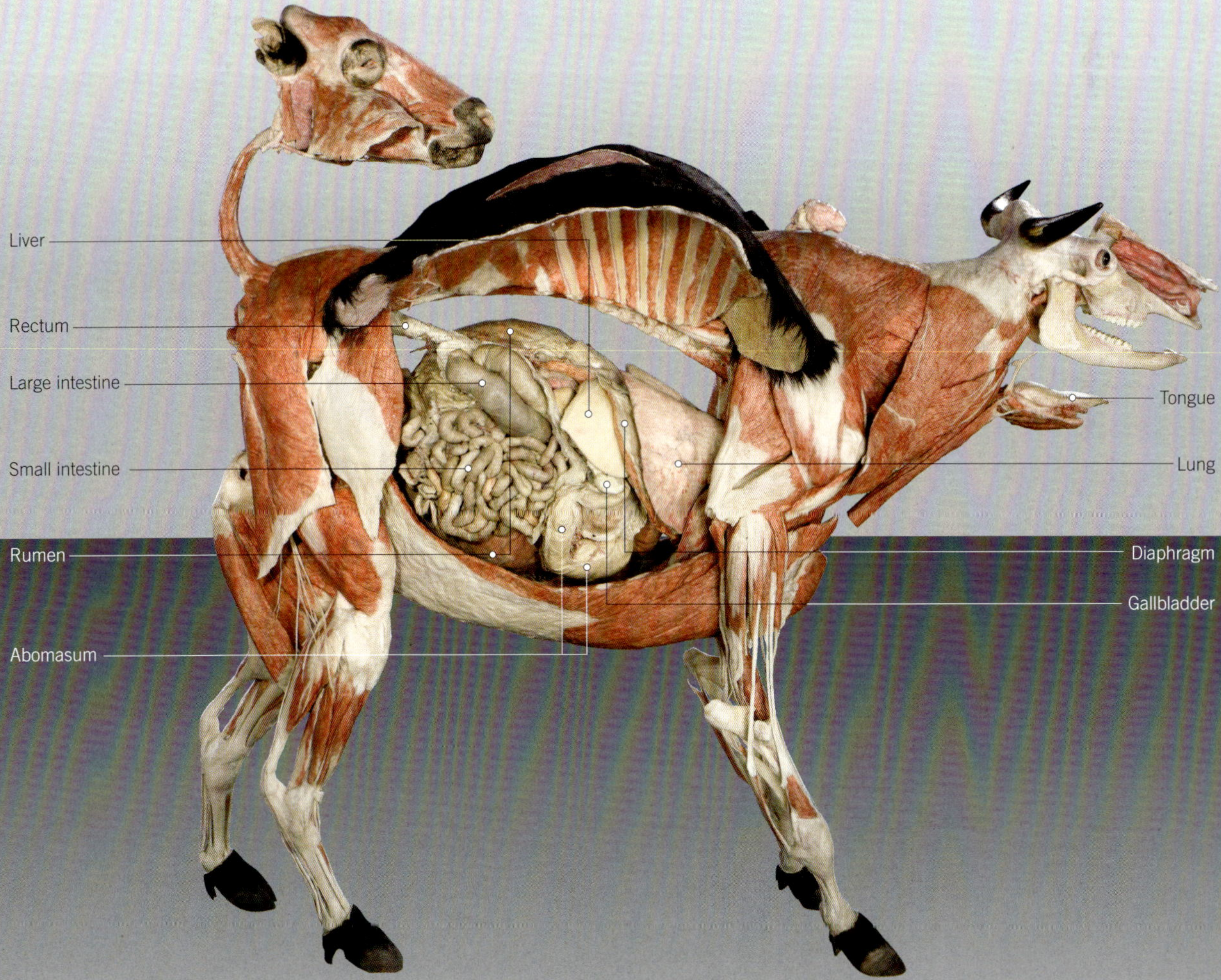

Yak, view from the right side
(Bos grunniens)

Filtering Waste

Every organism produces waste inside its body as a by-product of feeding and expending energy. Filtering blood is a common method in the animal kingdom for excreting such wastes.

In vertebrates,
the kidneys fulfil this vital function.
Their special micro filters continuously extract dissolved waste products and water from the blood and send them to the bladder as urine, to be expelled.
At the same time, the kidneys maintain the internal balance of liquids, minerals and trace elements in the body.

Often urine also contains hormones or other substances that serve communication purposes in some animals, for example signalling a readiness to mate.

Many animals spray urine
to mark their territory.

Urinary Tract

The entire blood volume
flows through the kidneys several times per hour
to filter out toxins and excrete substances
from the blood.

In general,
the kidneys consist of urine-producing
and urine-transporting structures:
the renal cortex, the medulla and the renal pelvis.
The renal cortex contains
a variety of small blood vessel clusters (glomeruli)
in which the urine is filtered out from the blood.
A pipe system transports the urine to the inner medulla.
Here the urine is concentrated
and finally passes through fine collection ducts
into the renal calices and further into the renal pelvis.
From here,
the urine flows down the ureters into the bladder.
When the bladder is full,
the urine is excreted through the urethra.

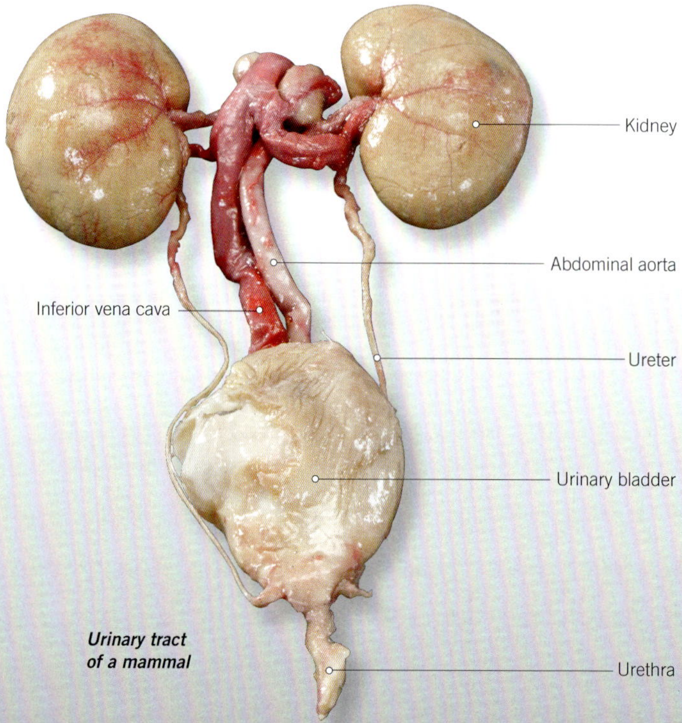

Urinary tract of a mammal

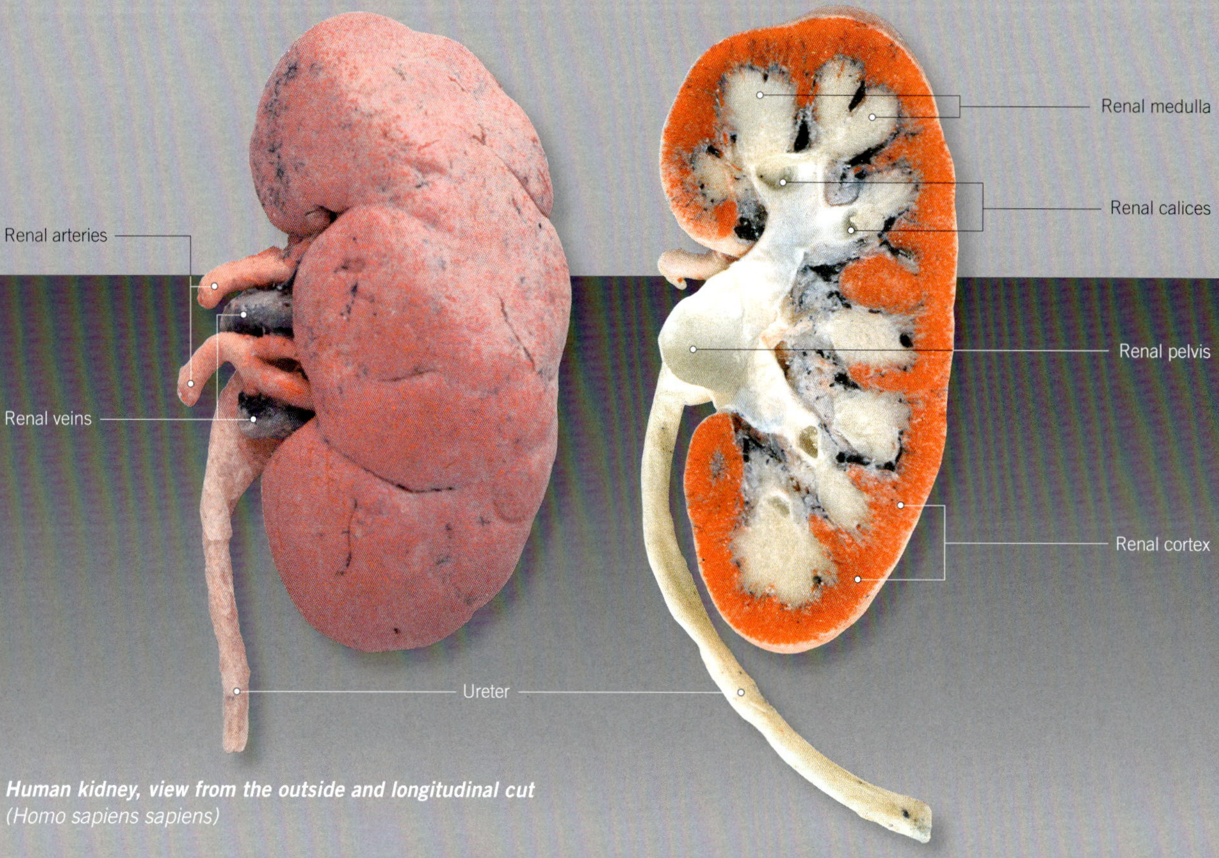

Human kidney, view from the outside and longitudinal cut
(Homo sapiens sapiens)

Reptiles and amphibians excrete urine from the anus.
The ureter, as well as the ducts of the sex organs,
lead into a section of the rectum, the *cloaca*.
The urine is concentrated by water reabsorption in the rectum
and therefore has a pasty consistency.

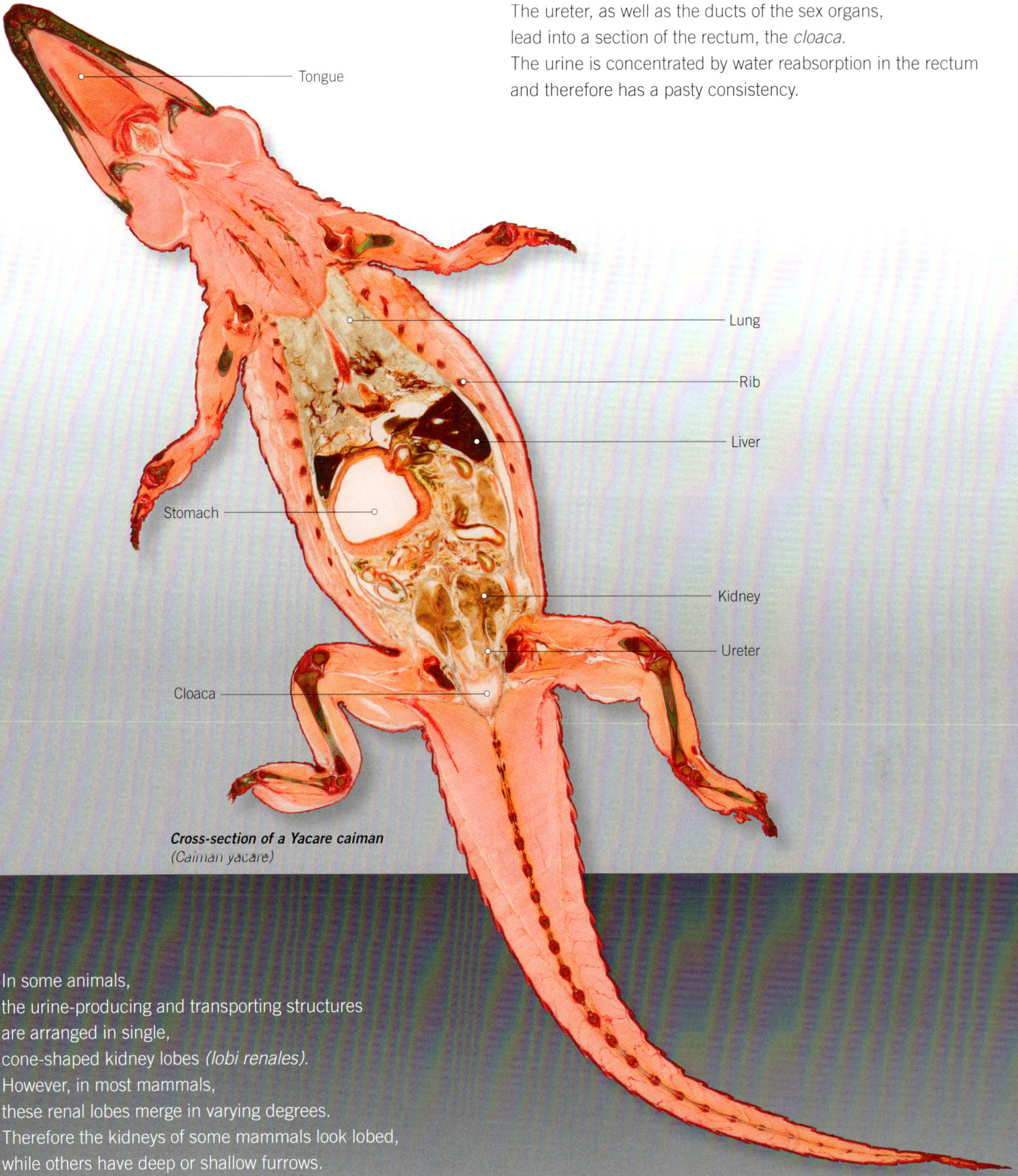

Cross-section of a Yacare caiman
(Caiman yacare)

In some animals,
the urine-producing and transporting structures
are arranged in single,
cone-shaped kidney lobes *(lobi renales)*.
However, in most mammals,
these renal lobes merge in varying degrees.
Therefore the kidneys of some mammals look lobed,
while others have deep or shallow furrows.
Still others have a completely smooth surface,
including the kidneys of pigs and humans.

Mainspring of Life

In rhinoceroses,
mating begins with foreplay lasting several days.
When the cow is ready to mate,
she walks backwards towards the bull.
The mating process itself lasts about half an hour.

Reproduction is a typical feature of all living beings.
In addition to feeding,
it is the most important activity
to ensure survival of the species.

Over the course of evolution,
different forms of reproduction have developed
among animals.

In mammals, sexual, or gamic, reproduction
is the most common form.
During the process, the father's semen (sperm)
fuses with one or more of the mother's eggs (ova).
In some animals, fertilisation happens outside the body.
In others, the sperm is deposited into the female's body
with the help of a special sexual organ.

Every fertilised ovum contains the entire blueprint
for the new being, the genome.
It is made up of maternal and paternal paired genes
that are arranged in chromosomes.

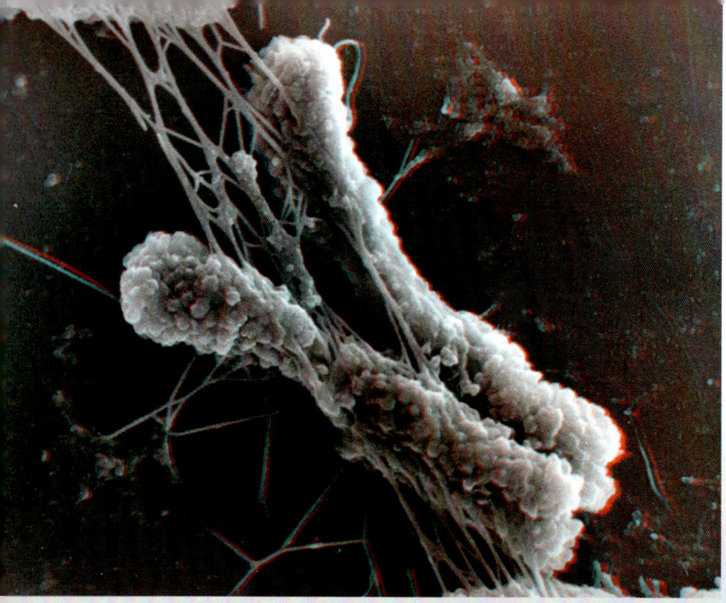

Genes not only determine
the individual properties of a new being,
such as its body size, the colour of its fur or
its behavioural patterns,
but also its resistance against pathogens
or other environmental factors.

By mixing the genetic material,
the resulting offspring differ from each other
and their parents.

This genetic variety enables animals
to adapt to changing environmental conditions
over generations.
This adaptability is a significant contribution
to the survival of a species.

Stereoscopic image of a chromosome. Chromosomes are thread-like structures inside the nucleus. They contain the genes that carry all hereditary information.

The female reproductive organs of a human being
(Homo sapiens sapiens)

Female Reproductive Organs

All young mammals develop in the protected environment
of their mother's womb and are fed milk after birth.

In female mammals,
the reproductive organs are mostly hidden inside the body.
These include two ovaries, Fallopian tubes, the uterus
and the vagina.

All of the female gametes, the ova,
are already present in the ovaries at birth.
They mature in cycles,
the length of which varies for different animals.

The matured oocytes are released into the Fallopian tube,
where they will merge with the sperm.
If the egg is fertilised, an embryo will develop.

A fertilised egg cell migrates into the uterus,
nests in the *endometrium*, or lining, and develops into a foetus.
It is protected in the womb.
When all the organs are developed
and the foetus has reached a certain size, birth begins.
The contracting muscles of the uterus
force the offspring out through the vagina.

Reindeer foetus inside the uterus
(Rangifer tarandus tarandus)

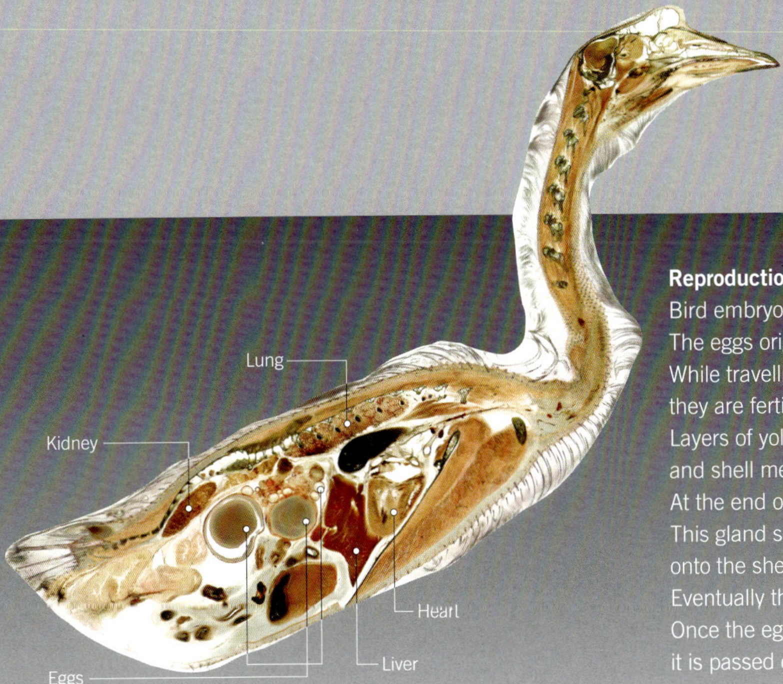

Longitudinal section of a goose with maturing eggs in the oviduct
(Anserinae)

Reproduction of Birds
Bird embryos mature in eggs largely outside the body cavity.
The eggs originate as tiny ova inside the ovary.
While travelling down the oviduct,
they are fertilised by male sperm.
Layers of yolk and albumen are added,
and shell membranes form.
At the end of the oviduct, there is a shell gland.
This gland secretes a viscous lime-containing mass
onto the shell membranes.
Eventually this mass hardens, forming the eggshell.
Once the egg is fully formed,
it is passed out of the body through the *cloaca*,
the joint excretory duct
of the reproductive and digestive organs.

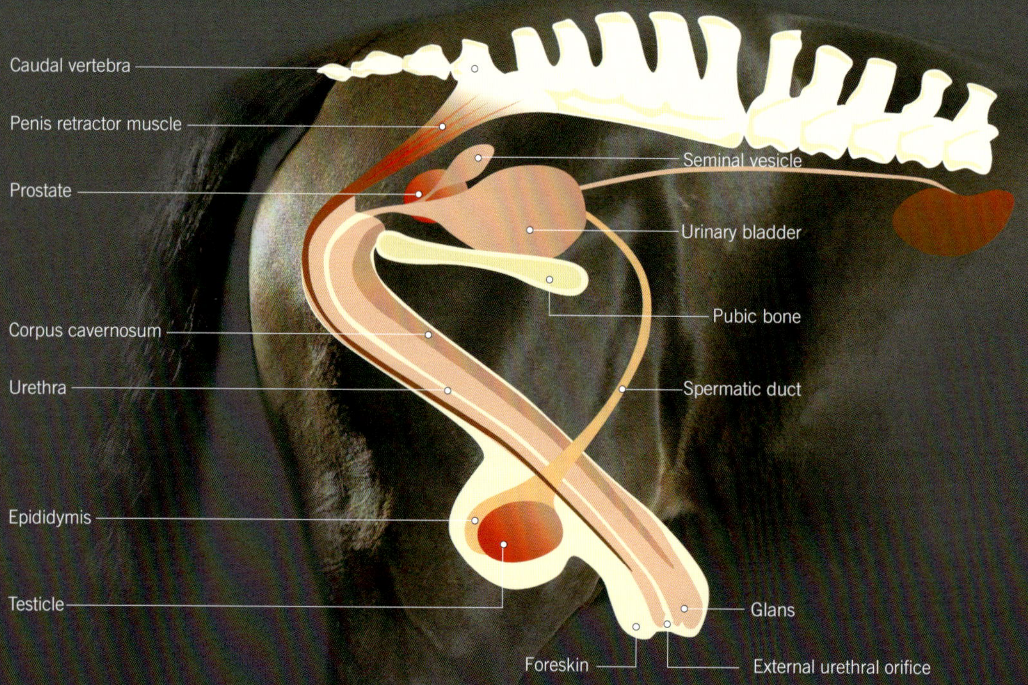

Schematic illustration of a stallion's reproductive organs

Male Reproductive Organs

In male mammals, the penis and scrotum make up the external reproductive organs. The scrotum holds two egg-shaped testicles. They produce the male gametes, the sperm. Sperm are stored in the *epididymides*, long coiled tubes that envelope the testes like a hood. The internal reproductive organs also include the seminal vesicles and the prostate gland.

Both add liquids to the seminal fluid at the moment of ejaculation. These liquids are important for the mobility and viability of the sperm travelling towards an ovum.

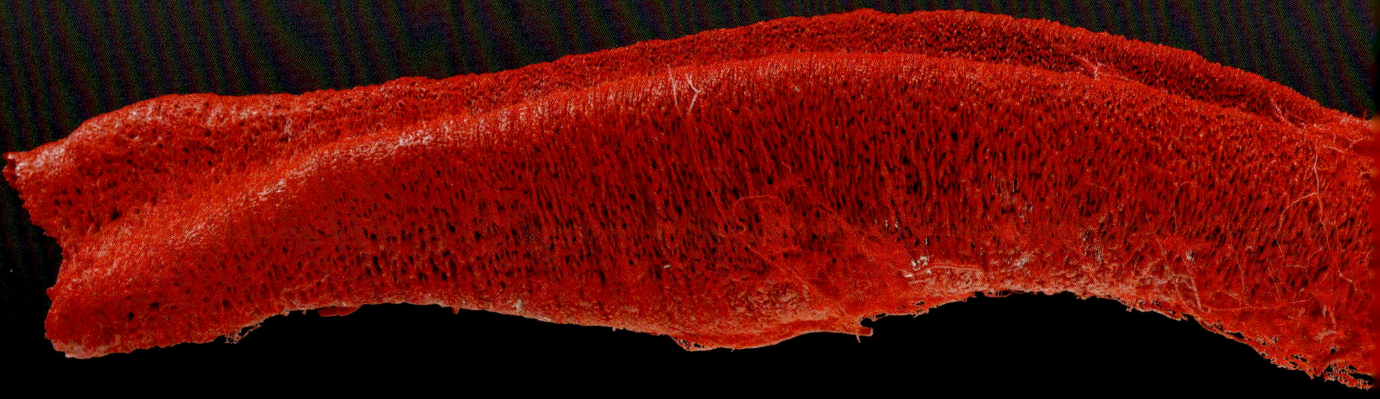

Blood vessel configuration of a horse's penis
(Equus ferus caballus)

The testicles contain thousands of finely coiled tubes, called *seminiferous tubules*, in which several hundreds of millions of sperm cells mature every day.

Unlike the gametes of female mammals, the gametes of males are located outside the body. It would get too hot for them inside the abdominal cavity.

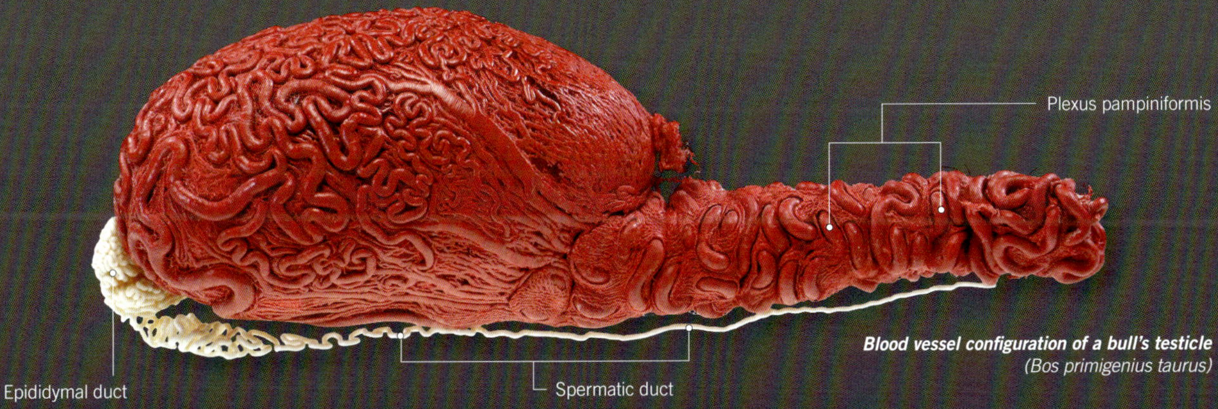

Blood vessel configuration of a bull's testicle
(Bos primigenius taurus)

A mammal's penis contains three sponge-like cavernous bodies that fill with blood when in a state of sexual arousal. This makes the penis larger and harder, resulting in an erection.

In some mammals, the penis contains a bone or a cartilaginous tube. This is the case in rodents, bats, baleen whales, simians and prosimians, as well as most marsupials and carnivorans.

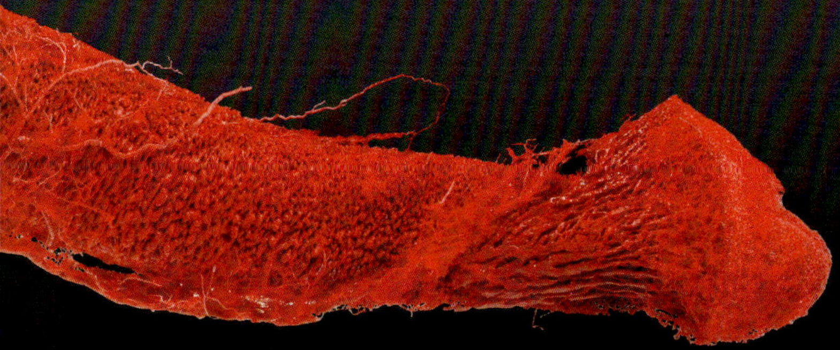

Whole-Body PLASTINATES

The Sea-tizen

Sharks inhabit almost all areas of the oceans and, with some exceptions, rivers and lakes. In order to survive in such diverse habitats, they have adapted perfectly to their environment. There are over 500 different shark species which vary tremendously in shape and size.

Mako shark
(Isurus)

Biologically,
sharks are part of the cartilaginous fish group (Chondrichthyes), along with rays and chimaeras.
That means that their skeleton is made of cartilage, in contrast to other fish that have bones.
Other major differences from most bony fish include:

- Sharks have no swim bladder.
- They have no gill plates, but gill slits.
- They have no scales, but dermal denticles.
- They reproduce by internal fertilisation and, depending on the species, give birth to only between 2 and 100 young – no comparison to the millions of eggs produced by bony fishes that are usually fertilised externally.

Mako shark
(Isurus)

The Senses of Sharks
Sharks have extremely well-developed sensory organs.

Their field of vision spans 320 degrees and they see better than cats in the dark.

Through their sensitive ears, they can determine the exact location of prey several miles away.

Their exceptionally good sense of smell allows them to detect blood in a concentration of 1:10 billion parts. This is like noticing a drop of wine in a medium sized swimming pool. In some shark species, the olfactory centre can even account for up to two-thirds of their brain mass.

A shark decides by the taste of its prey whether it is edible or not.

Like almost all fish, sharks have very sensitive sensors in the skin, the lateral line organs, through which they can perceive touch, water currents and changes in temperature.

RANGE OF SENSES

- **TASTE / TOUCH**
  Contact
- **ELECTRIC FIELDS**
  several centimetres / inches
- **PRESSURE**
  Lateral line (several metres/yards)
- **SIGHT**
  10 to 100 metres (11-110 yards)
- **SMELL**
  100 metres and more (110 yards and beyond)
- **HEARING**
  several kilometres/miles

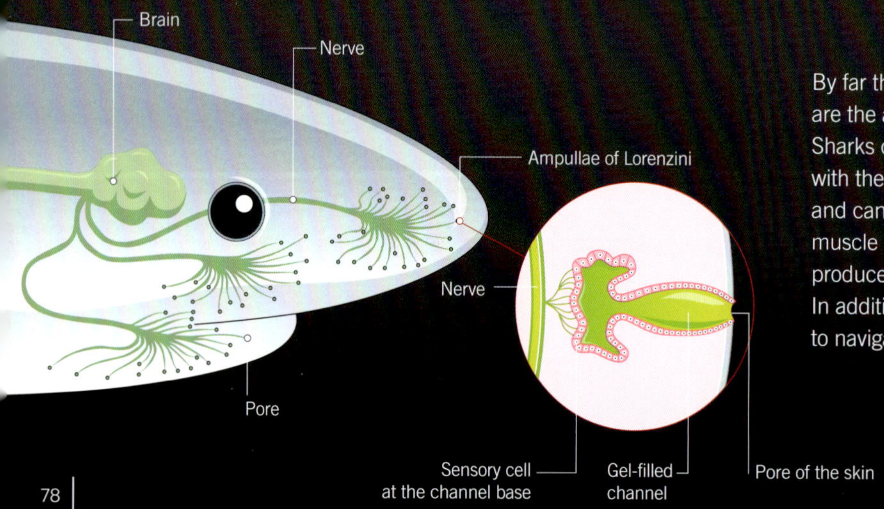

By far the most amazing sensory organ of sharks are the ampullae of Lorenzini. Sharks can sense even the weakest electric fields with these receptors and can therefore detect heartbeats, muscle movements or brain waves produced by other living things. In addition, sharks use the Earth's magnetic field to navigate.

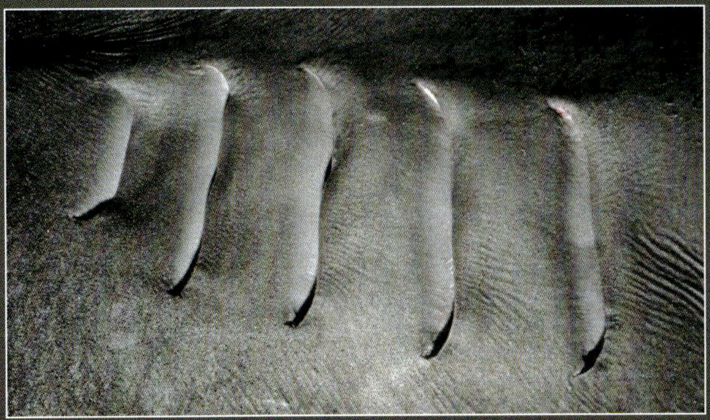

Gill slits

Most shark species can only breathe when they swim.
Their gill slits are not equipped with gill plates, found in other fish, which support the intake of breathing water.
A shark must constantly move to flush water through its gills and filter enough oxygen.
The water flows through the mouth into the gills and out again through the gill slits.

Mako shark
(Isurus)

Jack-of-all-Trades

Octopuses belong to the class of cephalopods
and rank among the oldest living things on Earth.
We distinguish about 200 different species.
Some are only a few centimetres/inches tall,
others have more than two-metre-long (6.5 ft long) tentacles.

Octopuses can be found in seas and oceans around the world,
and especially thrive in the Mediterranean Sea,
the Atlantic Ocean from Africa to Norway,
and the North Sea.
They usually live close to the sea floor
up to 200 metres (220 yds) deep and on rocky shores.

Octopuses have a sack-shaped body,
two large eyes and eight long arms.
Their name is derived from the number of arms they have.

Octopuses have no internal skeleton.
However, they are extremely robust and very flexible
and can squeeze even through the smallest openings.

Like all cephalopods,
the octopus breathes through gills
and cannot live on land permanently.
For brief periods of time, however,
the octopus is capable of leaving the water
and moving around on land.

Octopuses are particularly well known
for their highly developed nervous system.
They are very intelligent and quick learners.
For example, they can remember colours and forms,
find their way out of complex labyrinths,
and even open jars – skills we typically think
only mammals are capable of having.

Due to its great vision,
the octopus can adapt it's appearance to the colours,
brightness and texture of its surroundings within seconds.
It uses this skill for camouflage
and also for communication with other members of its species.

Its arms and the suction cups on them
are interspersed with particularly strong nerves.
Octopuses also have a favourite arm
that they use more frequently
than the other arms.

Octopus
(Octopus vulgaris)

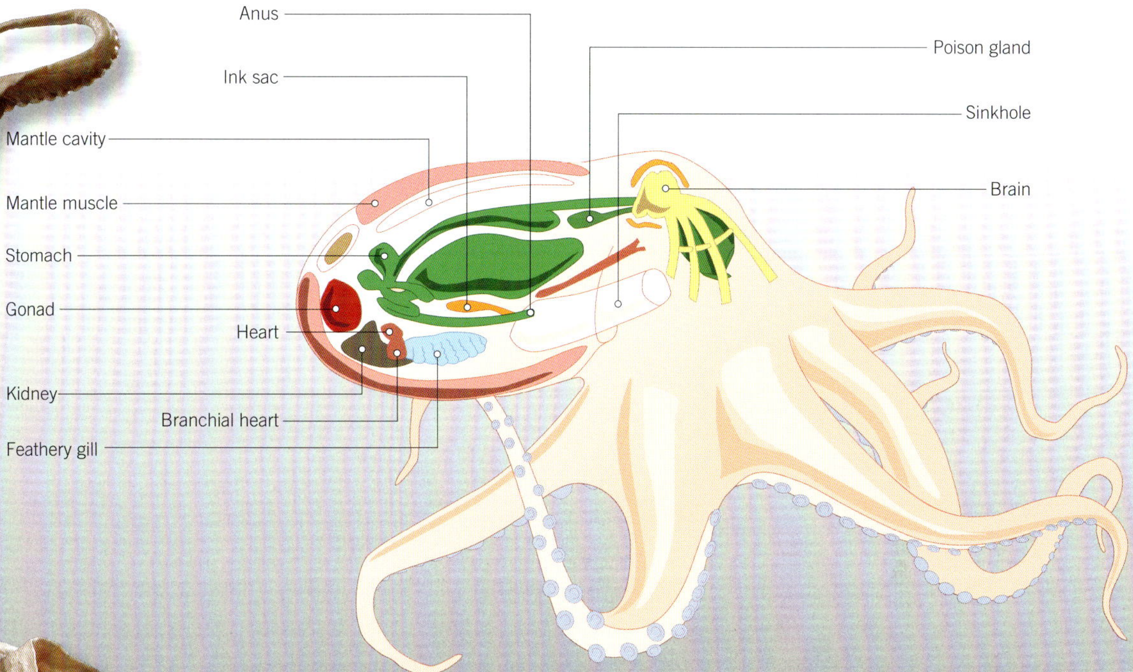

The oblong-oval shell encases all the internal organs. Attached to it are the head with the big eyes and the long tentacles. The mouth opening is at the bottom between all the arms.

Boneless
Wonder

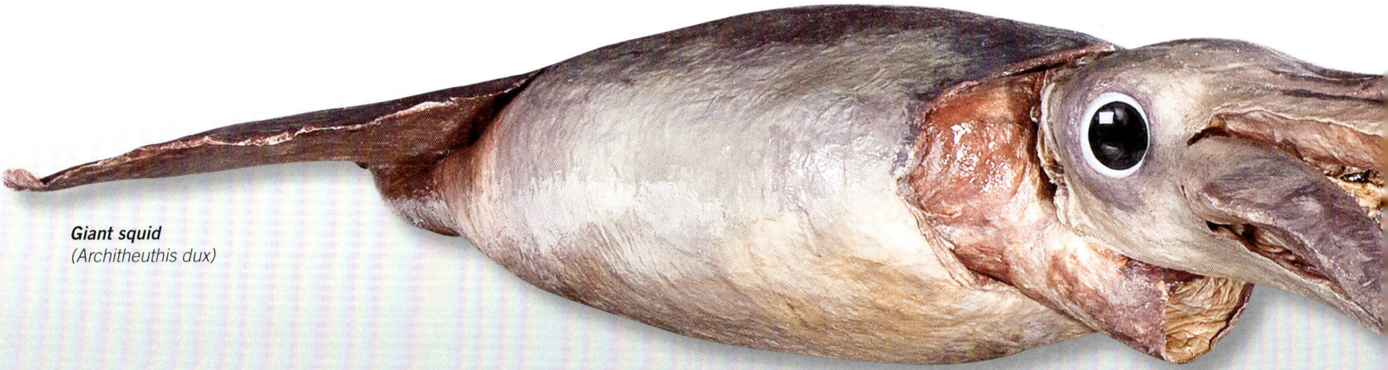

Giant squid
(Architheuthis dux)

Squid, like octopuses,
have tentacles attached to their heads
that they use to grasp their prey.
Thus, they also are called cephalopods –
"cephalo" meaning head and "pod" meaning foot.
The European squid of the North Atlantic Sea
can grow to be 30 to 45 centimetres (12–18 in) long.

Giant squid can reach lengths of up to 18 metres (20 yds)
and are the largest aquatic invertebrate organisms.
They live at depths of 300 to 600 metres (330–660 yds)
and are hunted by sperm whales.

Giant squid
(Architheuthis dux)

Squid have eight arms that are attached around the mouth opening. In addition, they have two retractable tentacles for grabbing prey.

Squid are the fastest swimmers among the invertebrates. They use jet propulsion as a specialised form of locomotion.

By relaxing their muscular mantle, they suck water into the mantle cavity. Powerful muscular contractions then squeeze the water through a funnel called a siphon. Thus, the cephalopod is catapulted forward. The siphon can also be rotated to control the direction of locomotion.

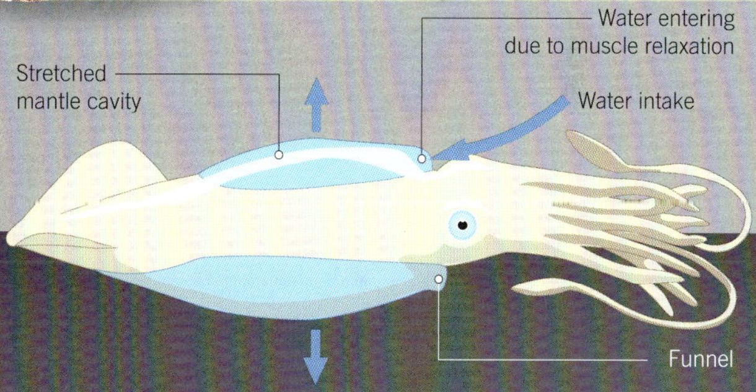

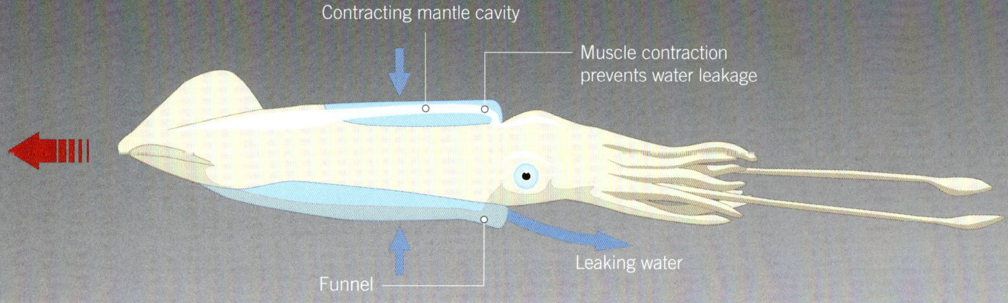

In Clouds of Ink

Like squid,
cuttlefish belong to the class of ten-armed cephalopods.
Their arms are relatively short.
At rest, the two longer tentacles
are usually hidden among the other arms.

Their mantle is generally more dull
and less wedge-shaped than that of the squid.
Another difference is their inner skeleton,
they have a cuttlebone that is made of carbonite mineral.
It contains a variety of gas-filled chambers
that give the animal its buoyancy.

Cuttlefish swim primarily using two fins
connected to the edge of the mantle,
which make wave-like movements for propulsion.
Therefore they are not quite as fast as squid.

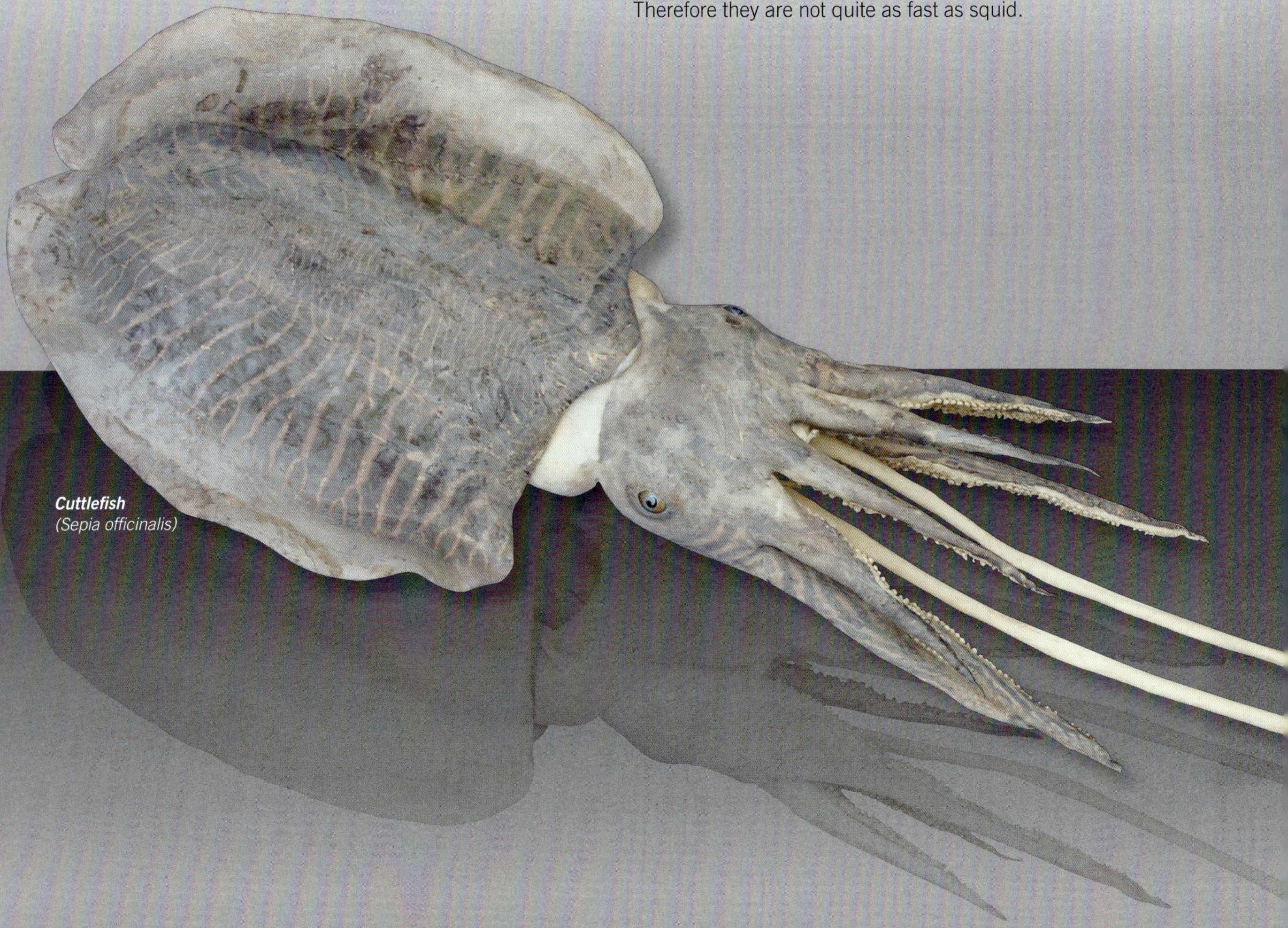

Cuttlefish
(Sepia officinalis)

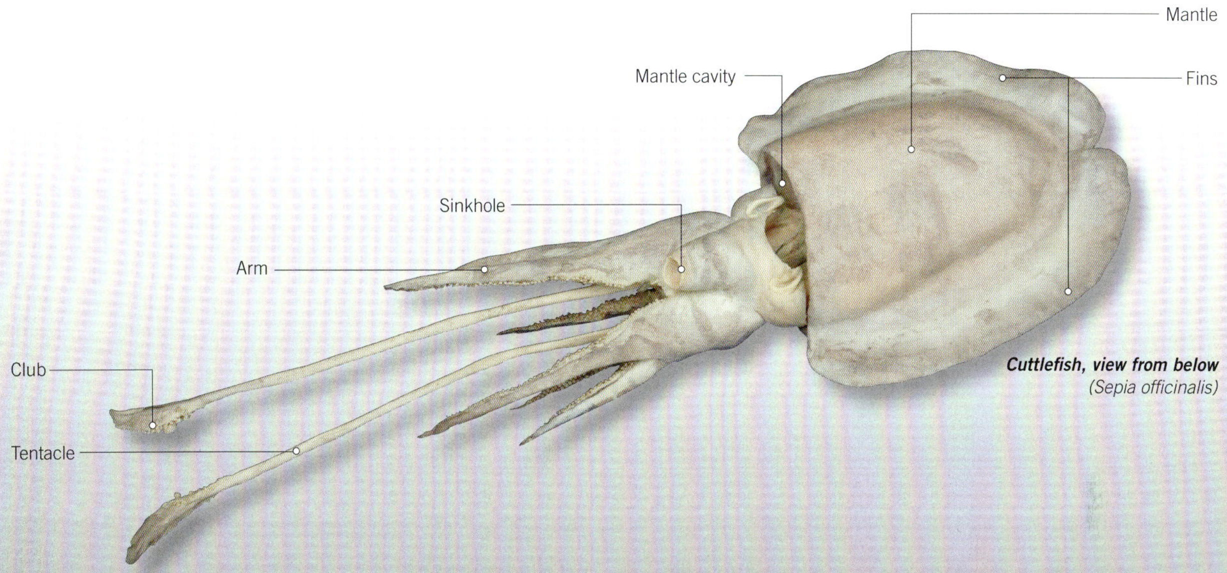

Cuttlefish, view from below
(Sepia officinalis)

Most cuttlefish are capable of changing colours and can bury themselves very quickly. With those skills they hide very well and do not have to escape far from predators.

The ink sac also serves to camouflage. It contains a dark ink of concentrated melanin. A cloud of ink obscures the view of the attacker and may suppress its sense of smell, while the cuttlefish escapes.

The Sweet-Tempered

Sheep were some of the first animals domesticated.
They were domesticated
from a different subspecies of wild sheep
probably as early as 9000 B.C.
They are robust and frugal,
which makes them adaptable
to a variety of climatic conditions and food availability.
That adaptability has certainly contributed
to large populations of these animals around the world.

Sheep
(Ovis aries)

Sheep are so biologically similar to goats that they are placed in the same tribe known as Caprini. Although their appearances are quite different, there is no significant difference in the internal structure of the organs, bones and muscles of sheep and goats.

The most important anatomical distinction between the two is the nature of their tails.
While sheep can only lift their tails up to a horizontal level, goats have a third muscle above their tails, allowing them to bring the tail into an upright position.

In many species of sheep, both males and females have horns.
They can be curved like a snail, long and spirally wound, or short and only slightly bent.

Sheep
(Ovis aries)

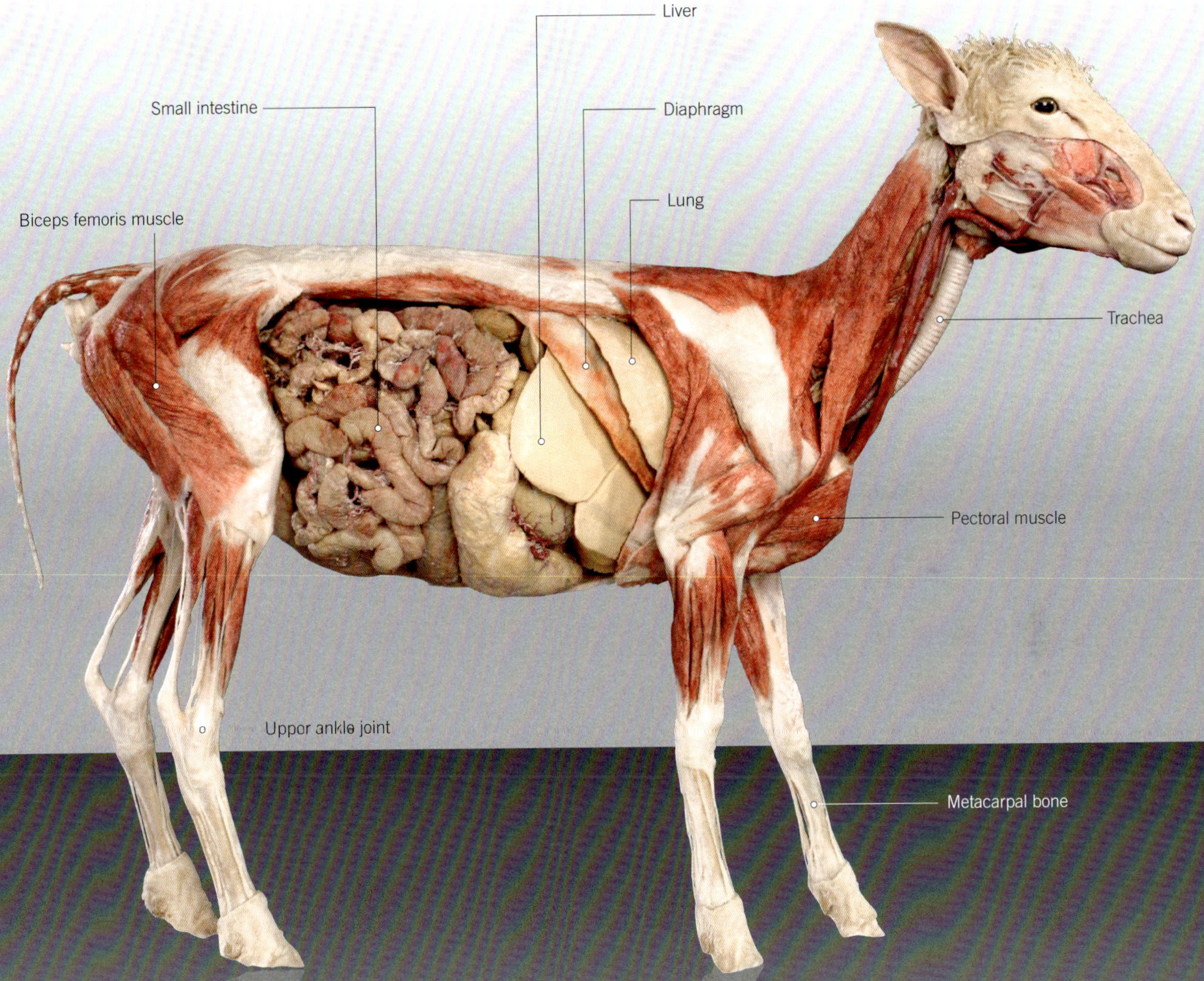

The term sheep generally refers to the female sheep.
They are also called ewes.
The males are referred to as rams
and are much larger and stronger than the female sheep.
Castrated sheep, or sterile males,
are called wethers.

They are much more peaceful than rams
and put on more meat.
Young sheep aged up to one year are referred to as lambs.

The Acrobats

Goat
(Capra aegagrus hircus)

In addition to sheep,
goats are some of the oldest domestic animals.
They grow up to 1 metre (3 ft) in shoulder height
and have a lifespan of 8 to 15 years.

Goats are even-toed unguligrades,
have short, thin legs, short fur, a short tail,
and horns that come in various shapes and sizes.
Both males and females have a characteristic goatee.

The goat's forelegs
are not connected to the chest by a joint,
but are attached only by muscles and ligaments.
Goats are very well coordinated,
which enables them to balance
on extremely uneven and rocky grounds.

Goats are ruminants.
Just like other ruminants,
they have no incisors in their upper jaws.
They break down food with their premolars, molars,
tongue and a tough pad on the hard palate.

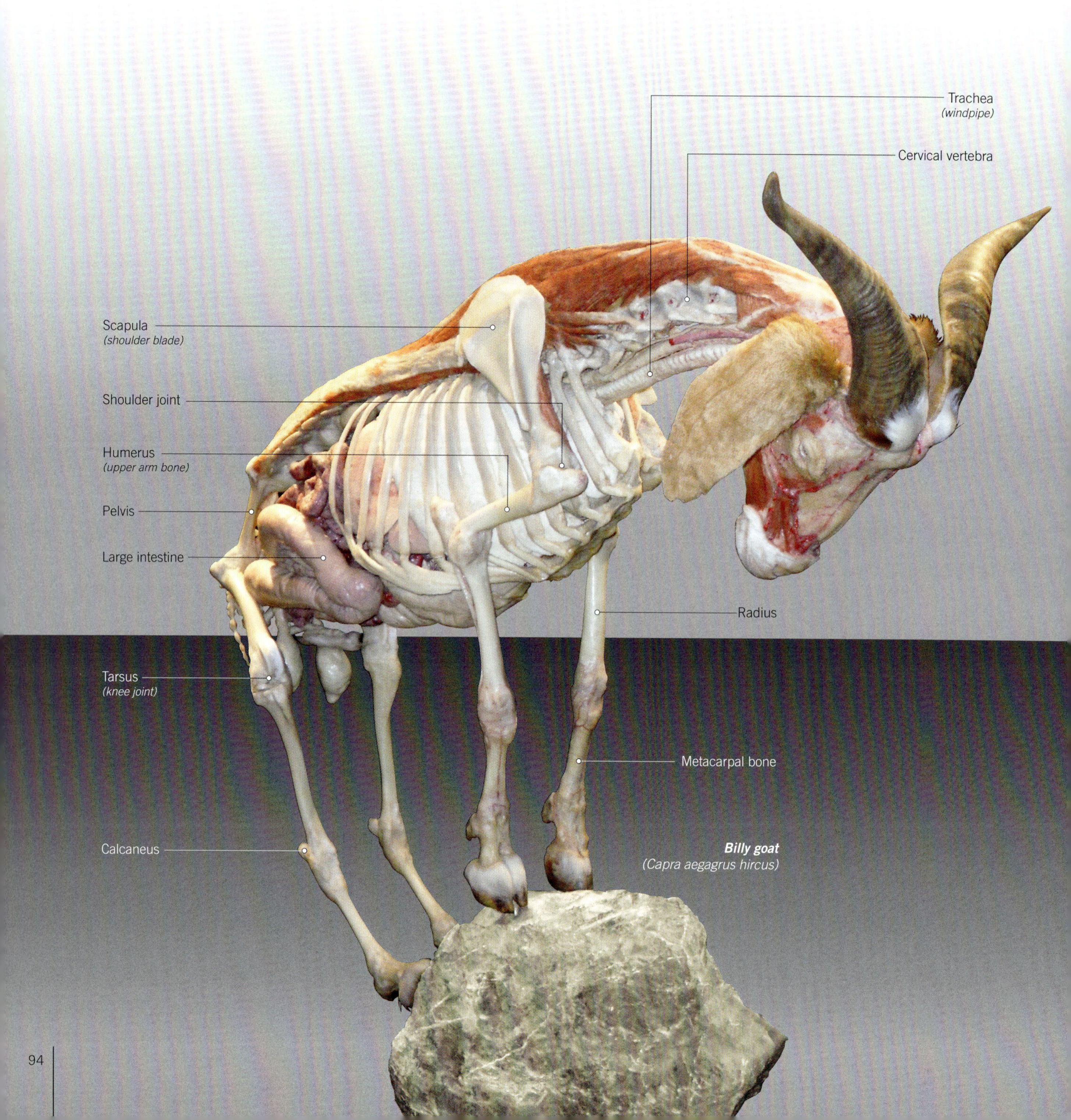

Billy goat
(Capra aegagrus hircus)

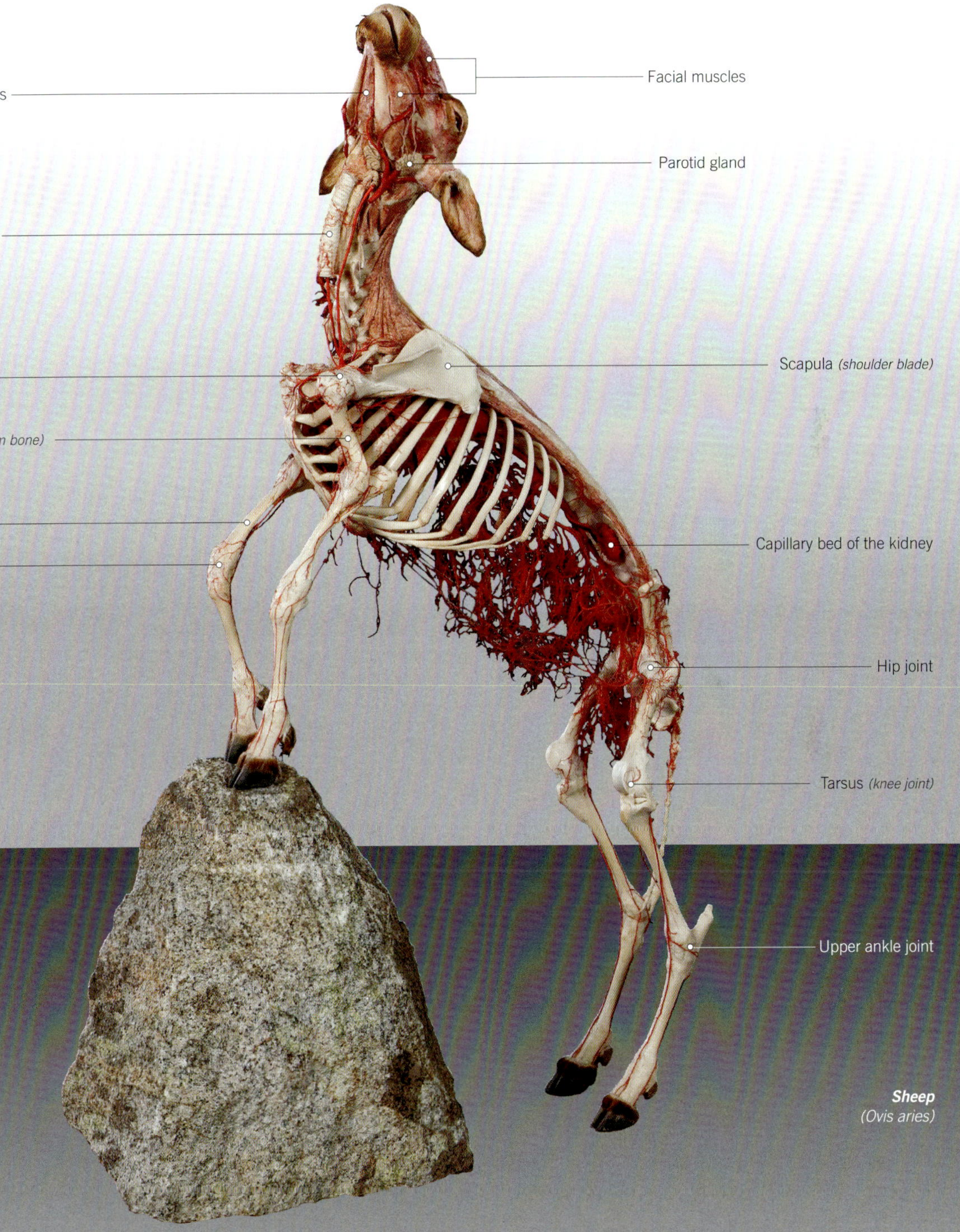

Sheep
(Ovis aries)

Gathered in Gallop

Before they were domesticated,
horses lived wild on the steppes.
As fast-running flight animals,
they are perfectly adapted to life in the grasslands.

Today there are a variety of horse breeds
that have been bred by humans.
They are divided into four groups,
according to their different temperaments.
Ponies make a fifth group.

Coldbloods are heavy, sturdy workhorses
with a massive head and strong neck.
They weigh up to 800 kilograms (1,764 lbs).

Trakehner and Hanoverian horses
are some of the most common warmbloods.
They are lighter and more agile than coldbloods.

Thoroughbreds have a relatively long body
with a narrow head and are very temperamental.

Half bloods are a cross-breed
between Thoroughbreds and a calmer class.
They are often used for various equestrian activities.

Foal
(Equus ferus caballus)

Horses have a characteristically long head.
Their eyes are big and located on the sides of the head, giving horses a full-surround view with the exception of dead angles right in front of the head and behind the animal.
Horses cannot see more than 2 metres (6.5 ft) directly in front of their heads, as their forehead essentially blocks their view.

Horses also have a fine sense of smell.
Their long noses offer a lot of room for the olfactory mucous membrane.
This allows them to detect smells from far away to find water and food and to avoid potential enemies.

The hearing capability of horses is about 200 times that of a human.
They can rotate their funnel-shaped ears independently of each other by 180 degrees.
This allows them to determine with great precision the direction from which a sound originates.
Hairs protect the ears against flies, dust and moisture.
If a horse folds its ears back, it is a sign that it feels uneasy or threatened.

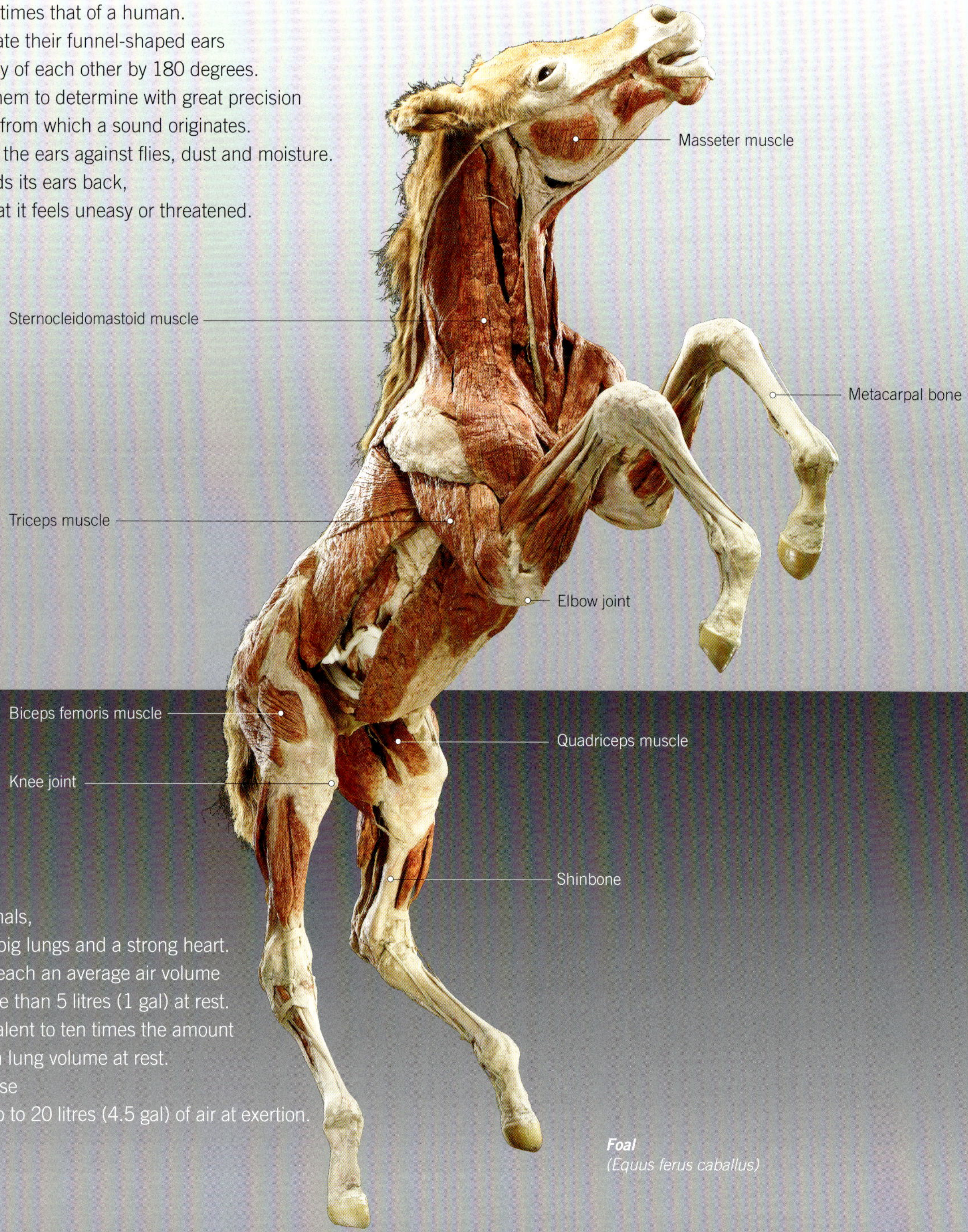

As flight animals, horses have big lungs and a strong heart.
Their lungs reach an average air volume of a little more than 5 litres (1 gal) at rest.
This is equivalent to ten times the amount of the human lung volume at rest.
A trained horse can inhale up to 20 litres (4.5 gal) of air at exertion.

Foal
(Equus ferus caballus)

As in all fast runners,
the humerus and the femur of the horse
are relatively short,
meaning that the elbows and knees are close to the torso.
The other arm and leg bones are relatively long,
especially those of the hands and feet.

Horses basically walk on their tiptoes.
The toes are grown back to the middle toe
and end in one toe that is enclosed by the hoof.
They are therefore also known as unguligrades.
The hoof is made of horn
and corresponds anatomically to the human finger nail.

Horses can sleep while standing up,
because they can lock their knee joints,
so that the muscles can rest.
However, they do not fall into deep sleep in this position.
For this, they need to lie down.

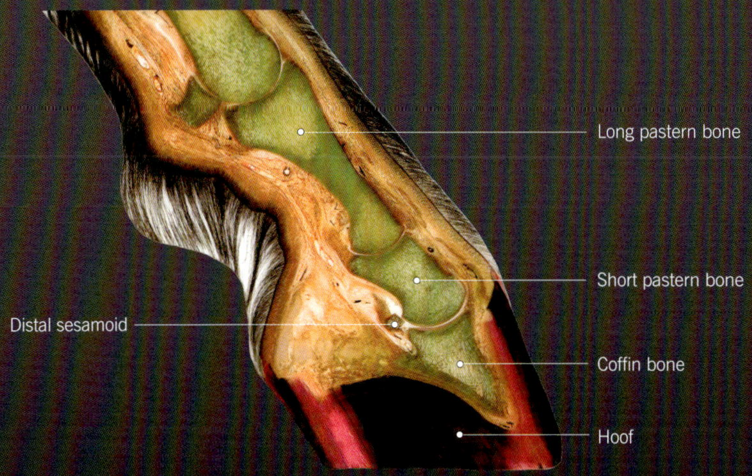

Longitudinal section of a horse's hoof
(Equus ferus caballus)

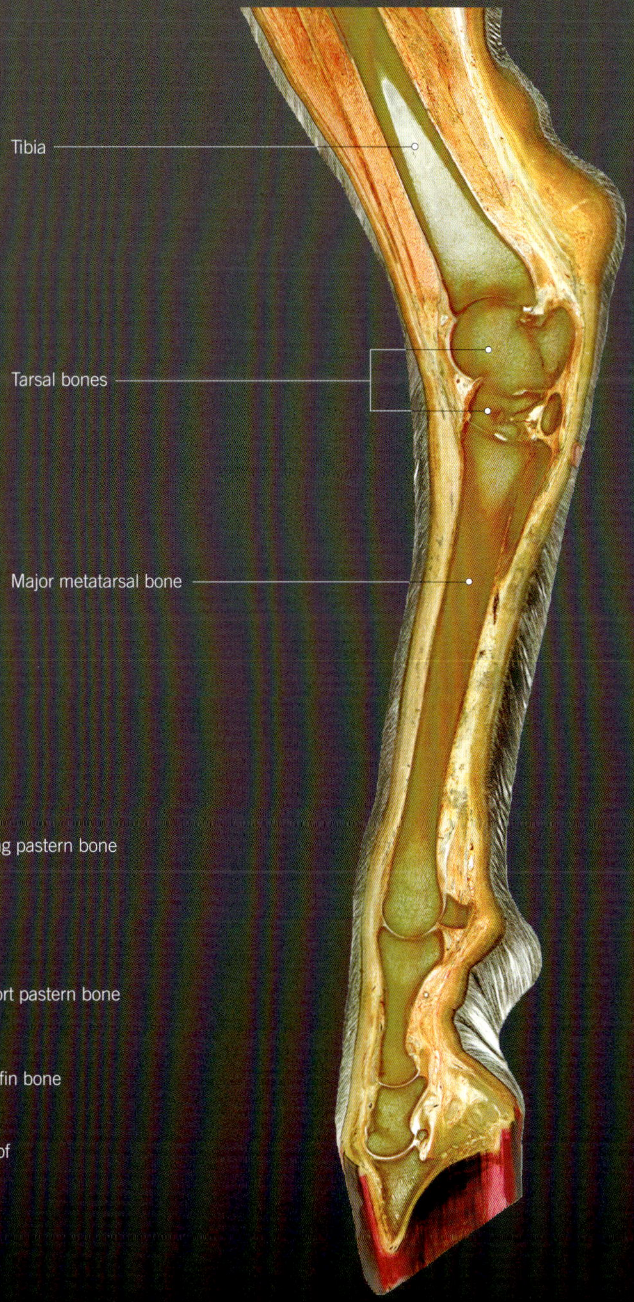

Longitudinal section of a horse's hind foot
(Equus ferus caballus)

The Restless

Reindeer belong to the family of deer.
They are among the most northerly living large mammals and are native to the Arctic regions of Canada, Alaska, Siberia and Scandinavia.
To escape the Arctic winter, reindeer herds travel long distances of up to 5,000 kilometres (3,100 miles).
Both males and females have antlers which they regularly shed and regrow.

Reindeer
(Rangifer tarandus tarandus)

Reindeer are ruminants
and feed exclusively on plant materials.
Their legs are comparatively long,
to the rest of their body,
as is the case with all fast runners.
This allows them to take large strides
and reach high speeds.
The muscles used to move their legs are inside their torso
to reduce the weight at the end of the legs.

Reindeer
(Rangifer tarandus tarandus)

Reindeer are *artiodactyls.*
That means that each foot
comes with two equally developed toes.
Evolutionally, these two toes
are the original third and fourth toe of the foot.
The first toe is completely absent,
and the second and fifth toe are much reduced
and do not even touch the ground.

The hooves of the reindeer
adapt to the changing seasons.
In summer, when the soil is moist and soft,
they are spongy and elastic.
In winter, they shrink and condense,
so that the hooves get sharp edges.
The edged hooves keep the animals from sinking
into snow and ice.

Male reindeer have an inflatable air sac
in front of the larynx,
enabling them to produce rattling coughing sounds.
This allows them to modulate their voice.
With a swelled neck and the characteristic rattling sound,
they determine the hierarchy among them
and compete for females during the breeding season.

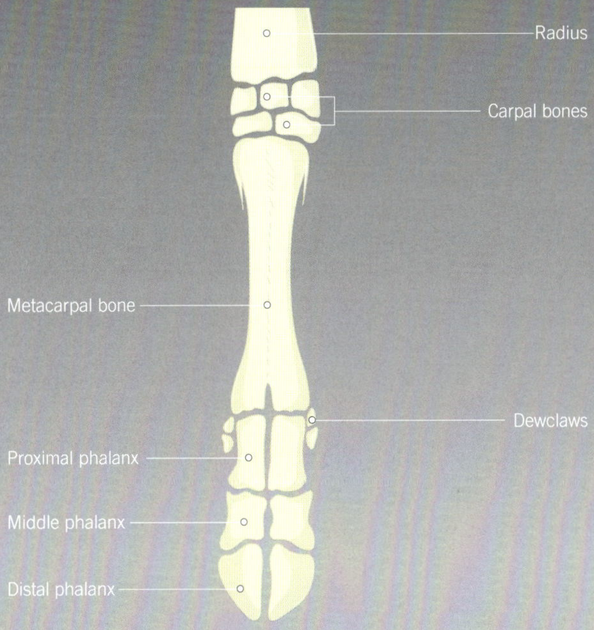

The Mountain Dweller

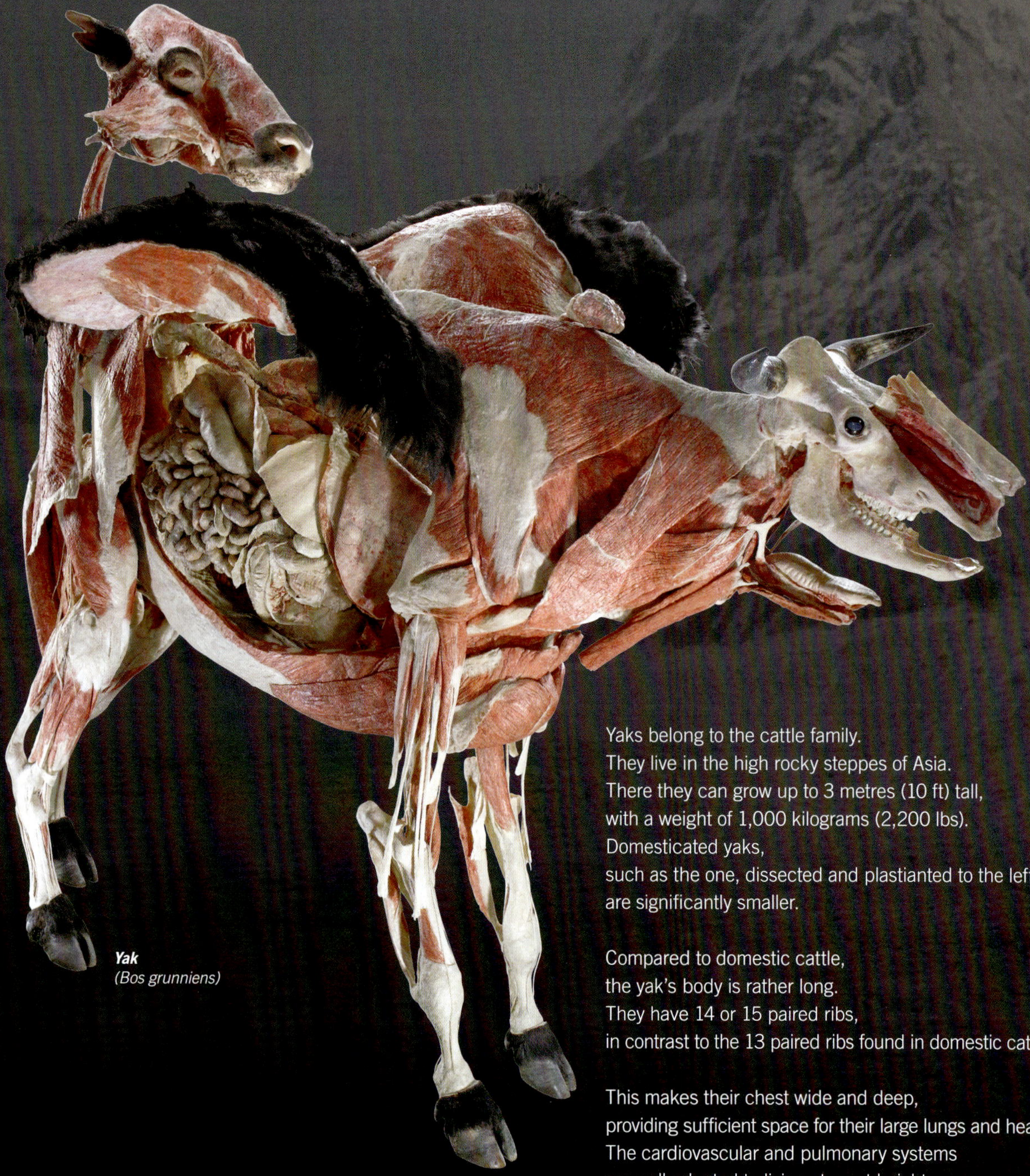

Yak
(Bos grunniens)

Yaks belong to the cattle family.
They live in the high rocky steppes of Asia.
There they can grow up to 3 metres (10 ft) tall,
with a weight of 1,000 kilograms (2,200 lbs).
Domesticated yaks,
such as the one, dissected and plastianted to the left,
are significantly smaller.

Compared to domestic cattle,
the yak's body is rather long.
They have 14 or 15 paired ribs,
in contrast to the 13 paired ribs found in domestic cattle.

This makes their chest wide and deep,
providing sufficient space for their large lungs and heart.
The cardiovascular and pulmonary systems
are well adapted to living at great heights.

In this specimen,
the abdominal wall was opened up and folded upwards,
giving the impression of a winged animal.
The muscles of the head were separated from the skull
and are shown separately.

Yak
(Bos grunniens)

Strong Like a Bear

Bears are considered majestic and powerful.
They have a stocky body, a large head,
and rather short and very strong extremities.

Depending on where they live,
they are either small or huge.
They can weigh 150 to 780 kilograms (330–1,700 lbs),
and their body size can reach 2 to 3 metres (6.5–10 ft)
in an upright position.

The smallest brown bears live in the Alps
and are roughly the same size as a St. Bernard.
The largest brown bears, the grizzly and the Kodiak,
are found in Asia and North America.
They are the biggest terrestrial predators on Earth.

Brown Bear
(Ursus arctos)

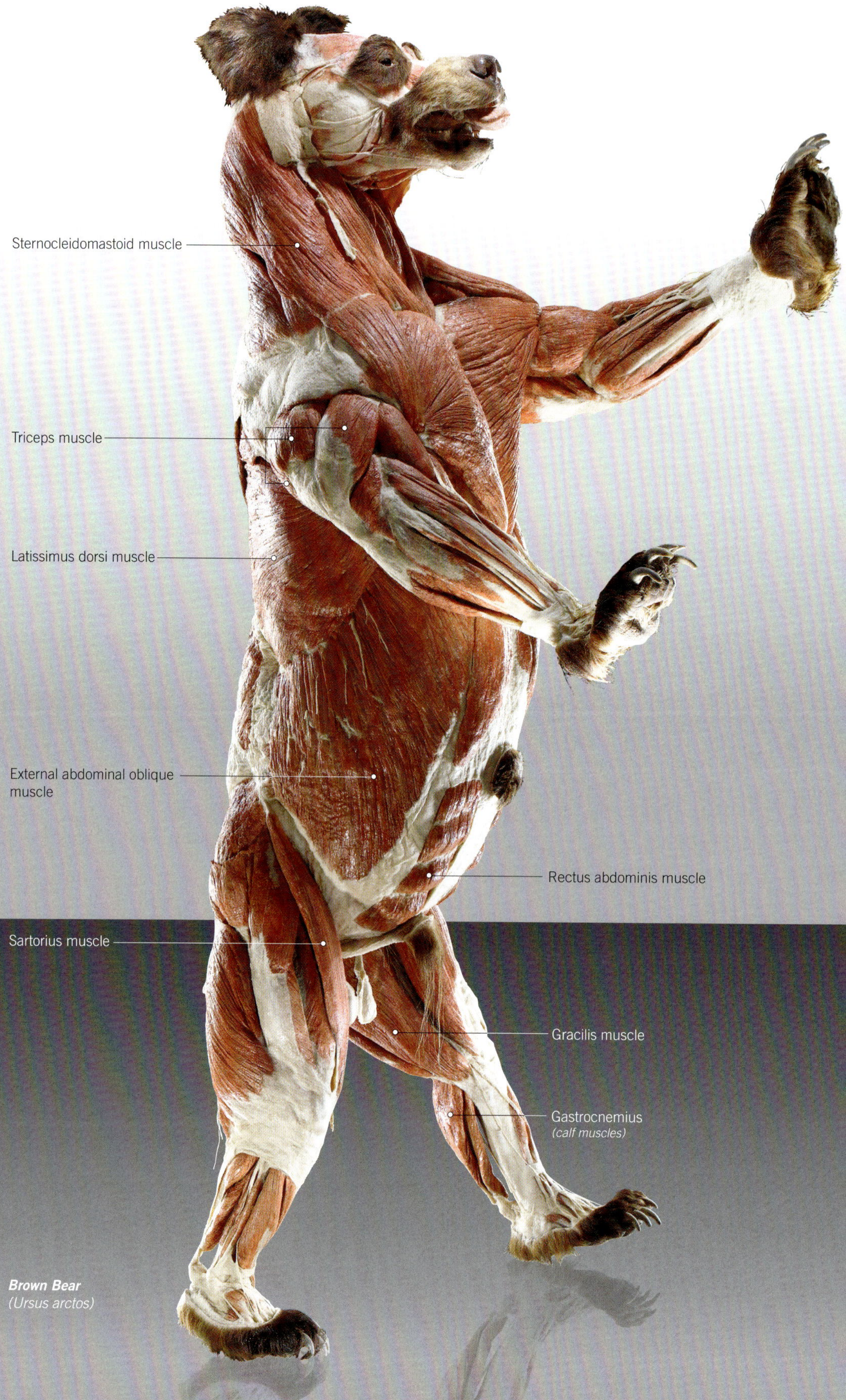

Brown Bear
(Ursus arctos)

Bears can move while upright,
but only for a short time.
When a bear rears up on its hind legs,
it does not do so to demonstrate its size or to attack,
but rather to get a better view.
A bear never attacks while up on two legs.

Bears amble, that is,
both legs of the same side of the body move simultaneously.
Usually, their movements are slow and sluggish,
but, if necessary, they can run very fast
and attain speeds of 55 km/h (35 mph).
They are also very good swimmers.

A bear's forelegs and hind legs
are almost equal in length,
ending in powerful paws,
each equipped with five non-retractable claws.
The soles of a bear's paws are cushioned
with fibrous connective tissue,
providing it with a soft gait on all fours.

Plantigrades, such as bears,
touch the ground with their heels,
metatarsals and toes.
Digitigrades, such as dogs,
touch the ground only with their toes.

The bones of a bear's forearm (ulna and radius)
and of its lower leg (tibia and fibula) are separate.
This allows the animal to make precise,
far-reaching movements in spite of its bulk.

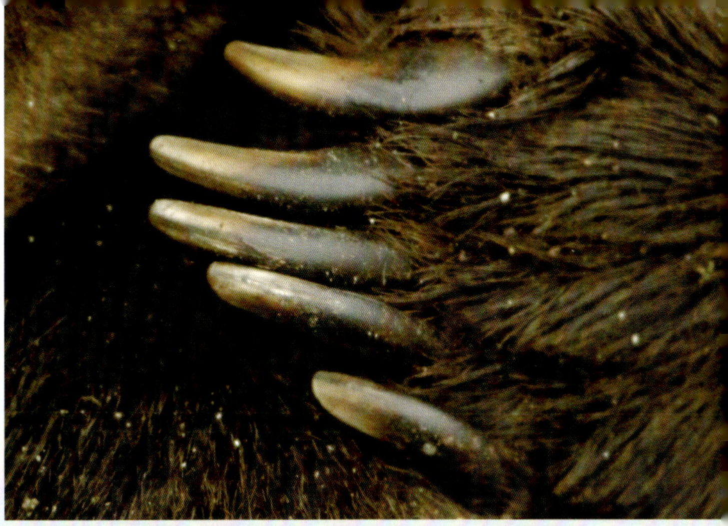

Claws of a brown bear
(Ursus arctos)

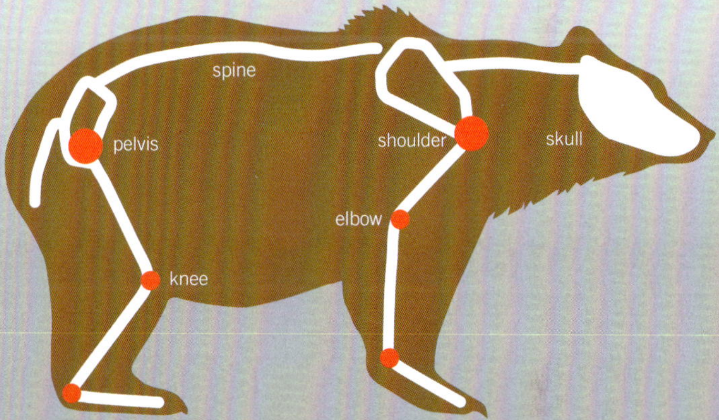

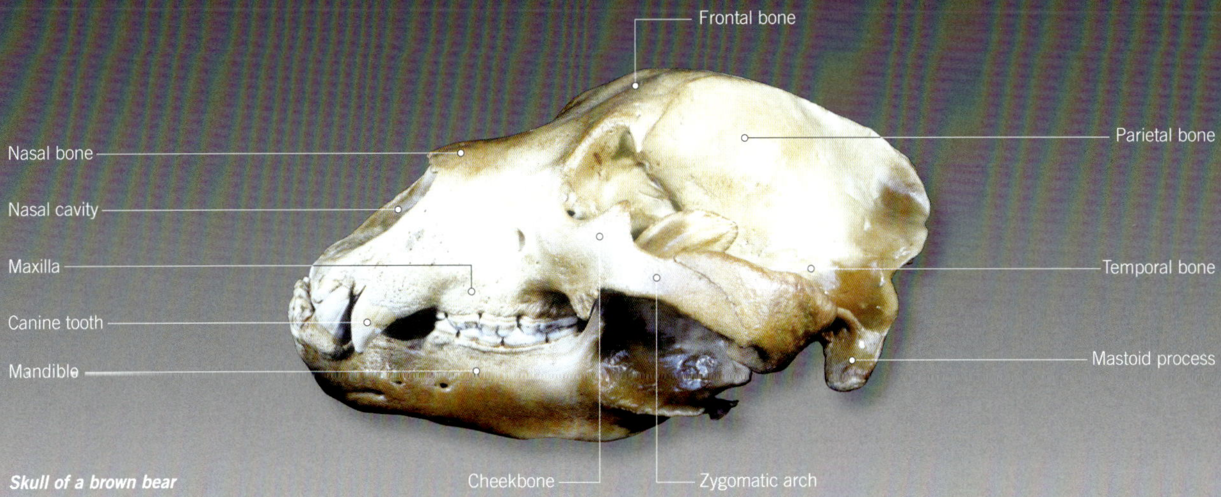

Skull of a brown bear
(Ursus arctos)

The Grounded One

Ostrich
(Struthio camelus)

The ostrich is the largest bird within the avian world.
It cannot fly
because it weighs too much to be able to lift off.
Ostriches can weigh up to 160 kilograms (353 lbs).

Birds usually have a bony protrusion on their breastbones,
the keel, where the flight muscles are attached.
In ostriches, however, the breastbone is rather flat,
and there are only poorly defined flight muscles.

On the other hand, ostriches are fast runners.
They reach speeds of up to 80 km/h (50 mph)
and are able to maintain them for about half an hour.
Very strong muscles in the back and legs facilitate this.
The long elastic ligaments of the ostrich act like springs.
They allow ostriches to regain energy
every time their feet push off the ground.

Unlike all other birds,
ostriches have only two toes –
an adaptation to their high-speed running.

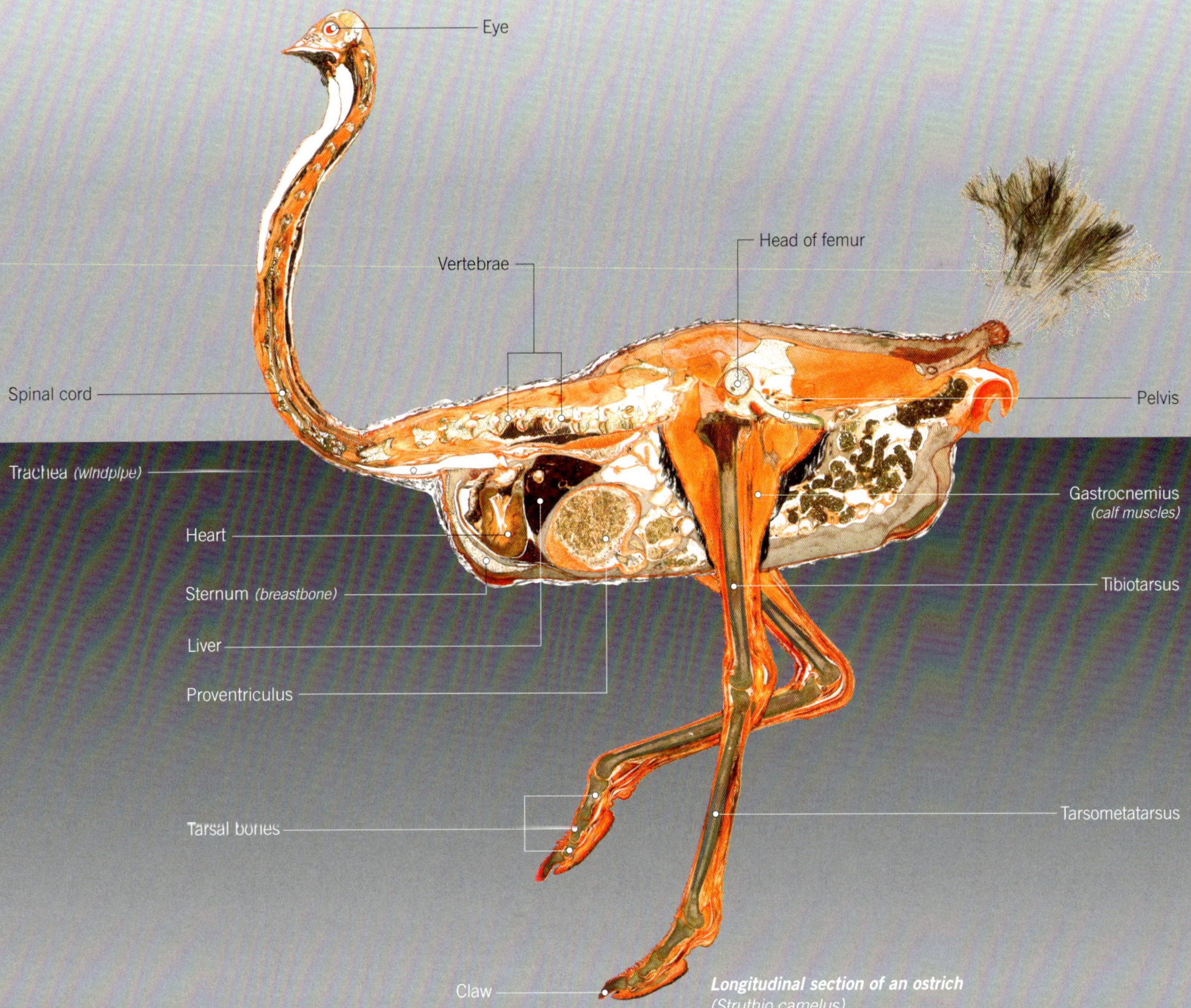

Longitudinal section of an ostrich
(Struthio camelus)

The Invincible

The Bactrian camel belongs to the genus Camelus.
It lives in the deserts and steppes of Asia
where it serves as a riding and pack animal.

With its two humps
it is easily distinguishable from the dromedary,
the one-humped camel.
The humps are not used for water storage,
as often assumed, but for fat storage.

Typical of camels is their rocking gait,
which is why they have the nickname “Desert Ship.”
This rocking gait, also called an amble, is caused from
the front and hind legs on the same side of the body
moving forward at the same time.

Bactrian camel
(Camelus bactrianus)

Bactrian camels live in very dry regions
with temperature fluctuations from -30°C to +40°C
(-22°F to +104°F).
Therefore they have developed a number of features
in order to adapt to their extreme living conditions.

Camels store water in the blood and stomach.
They can drink about 100 litres (26 gal) in a few minutes.
This enables them to survive for up to two weeks
without having to drink anything.
To save water and energy,
they excrete very concentrated urine and faeces.

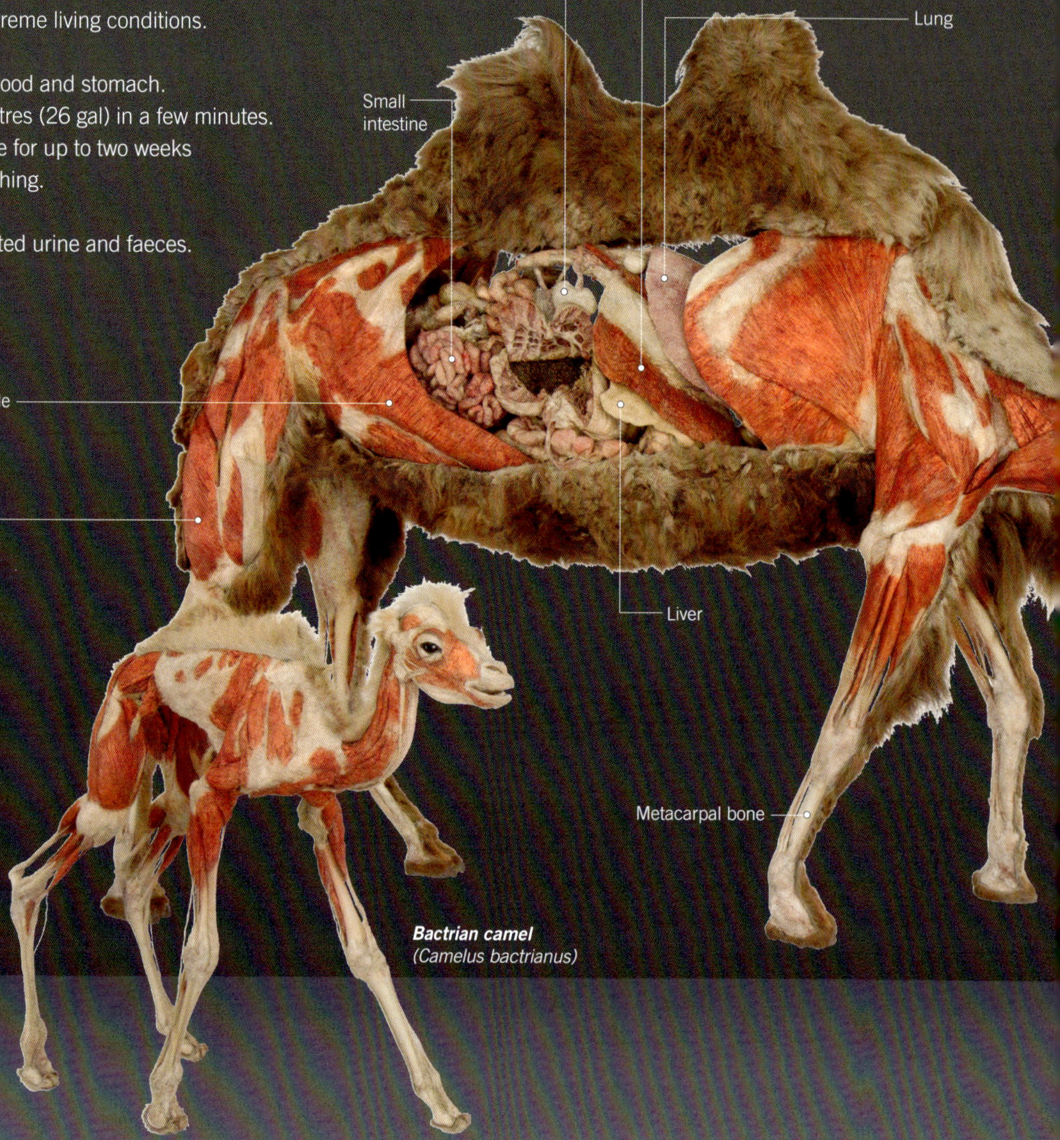

Bactrian camel
(Camelus bactrianus)

Their long eyelashes and sealable nostrils
protect their sensitive eyes and nose
from being penetrated by sand in a desert storm.

Their humps can contain
up to 40 kilograms (88 lbs) of fat as an energy reserve.
If the fat were evenly distributed throughout the body,
camels would quickly overheat.

Like all other artiodactyls,
Bactrian camels walk only on the digits
of the third and fourth toe.
However, they have no hooves
but bent nails that protect the front edge of the feet.
The toes rest on a cushion of connective tissue
which forms a broad leathery pad
and prevents the foot from sinking into the desert sand.
Camels therefore belong to the suborder of Tylopoda.

All camels are herbivores that regurgitate their food.
However, in a zoological sense
they do not belong to the ruminants
because they have evolved independently of that group.
In contrast to true ruminants,
two gastro-oesophageal vestibules equipped with glands
are located in front of the camel's main stomach.
These allow for the digestion of thorny, salty
and even poisonous plants.

After a gestation period of up to 440 days,
a single calve is born.
Camels rarely give birth to twins.
Newborn camels are *precocial*,
meaning they can walk just a few hours after birth.

Above the Rest

With its long neck,
the giraffe is the tallest living terrestrial animal.
Despite its height, it still has only seven cervical vertebrae
like most other mammals;
they are simply much longer.
One individual vertebra
can be up to 40 centimetres (16 in) long.

Giraffe
(Giraffa camelopardalis)

The blue-grey tongue of a giraffe can be up to 50 centimetres (20 in) long and serves as a grasping tool.

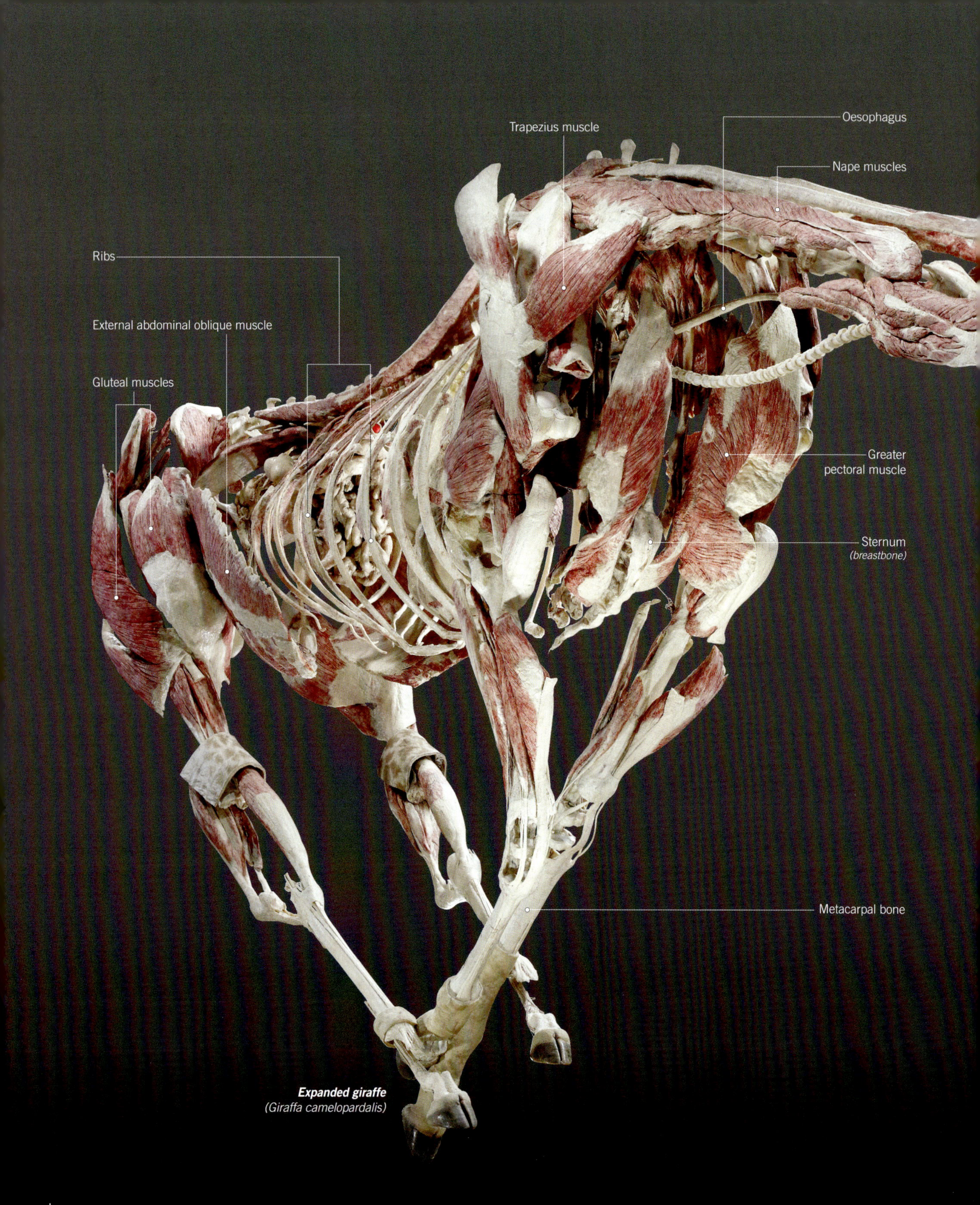

Expanded giraffe
(Giraffa camelopardalis)

Its long neck is an adaptation,
giving the giraffe an advantage over smaller herbivores.

It can reach up higher,
allowing it to pick the best parts of the highest trees.
Each day the animal consumes
more than 80 kilograms (176 lbs) of vegetable foods,
including thorny plants, such as acacia.
The inside of its cheeks
are lined with a thick layer of stiff skin cells
so that the thorns cannot penetrate.

Surprisingly,
giraffes are ruminants despite their long necks.
The pre-digested cud in the reticulum
is returned to the mouth
with the help of the muscular oesophagus.
They are the only animals
capable of walking while ruminating.
This is an adaptation to their nomadic way of life.

A single, extremely strong ligament supports the neck.
This ligament extends from the back of the head
to the coccyx
and is responsible for the "bump"
forming between the neck and the body.

Giraffes are distantly related to deer,
which can be recognised by their antler-like horns.
Both sexes have two to five bony horns on their forehead,
covered with skin.
In adult animals, they have grown attached to the skull.

When attacked, giraffes defend themselves
with kicks of their plate-sized hooves.
These kicks are lethal for the attackers.
A targeted kick can crush the skull of a predatory cat.
Therefore, in most cases
it is only old animals and young calves
that fall prey to an attack.
Giraffes die much more frequently of hunger
or the consequences of a fall.

Giraffes communicate outside a human's audible range
with frequencies below 20 Hertz,
the so-called infrasound range.

The giraffe's extremely long neck
is also a challenge to its cardiovascular system.
Its heart is particularly powerful,
enabling it to pump the blood up to the brain
against the force of gravity.
The giraffe's heart weighs about 10 kilograms (about 20 lbs),
and it can pump 60 litres (16 gal.) of blood per minute
through the body,
creating a blood pressure
that is three times that of humans.

The arteries in the giraffe's neck
have special valves making sure
that the giraffe will not suffer from brain damage
when suddenly lowering its head,
thereby significantly increasing the blood pressure
in its neck and brain vessels.

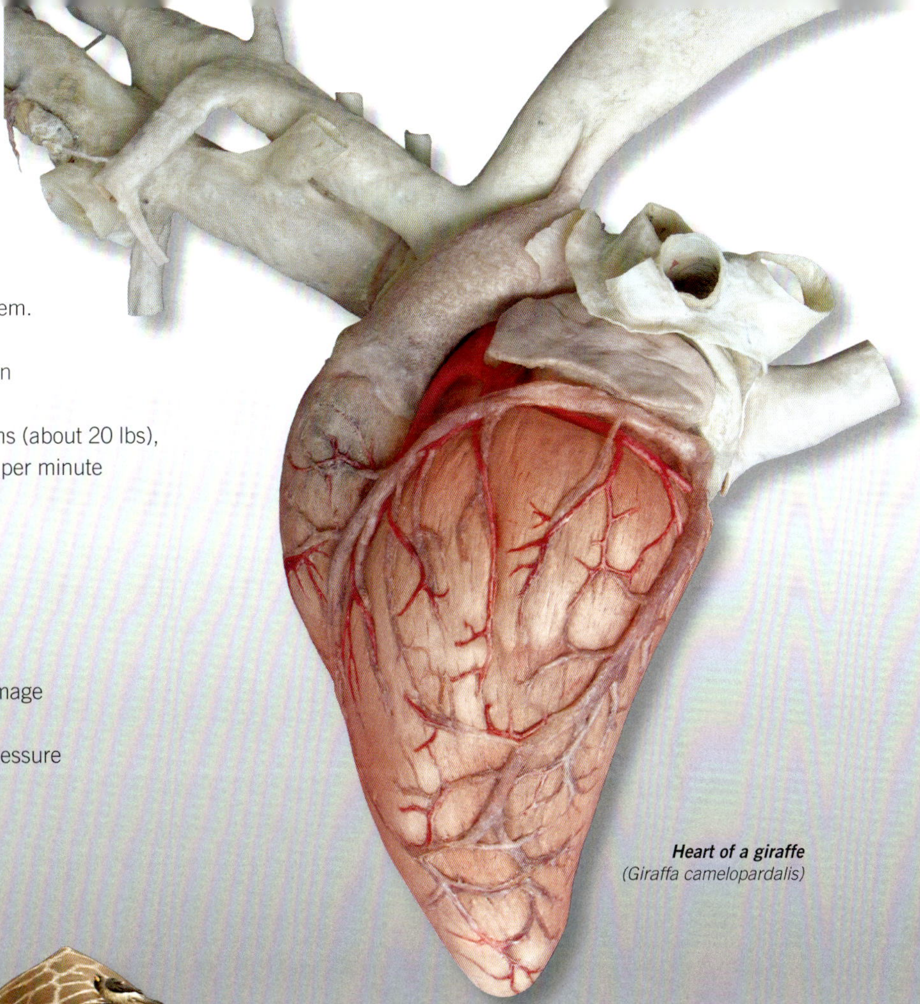

Heart of a giraffe
(Giraffa camelopardalis)

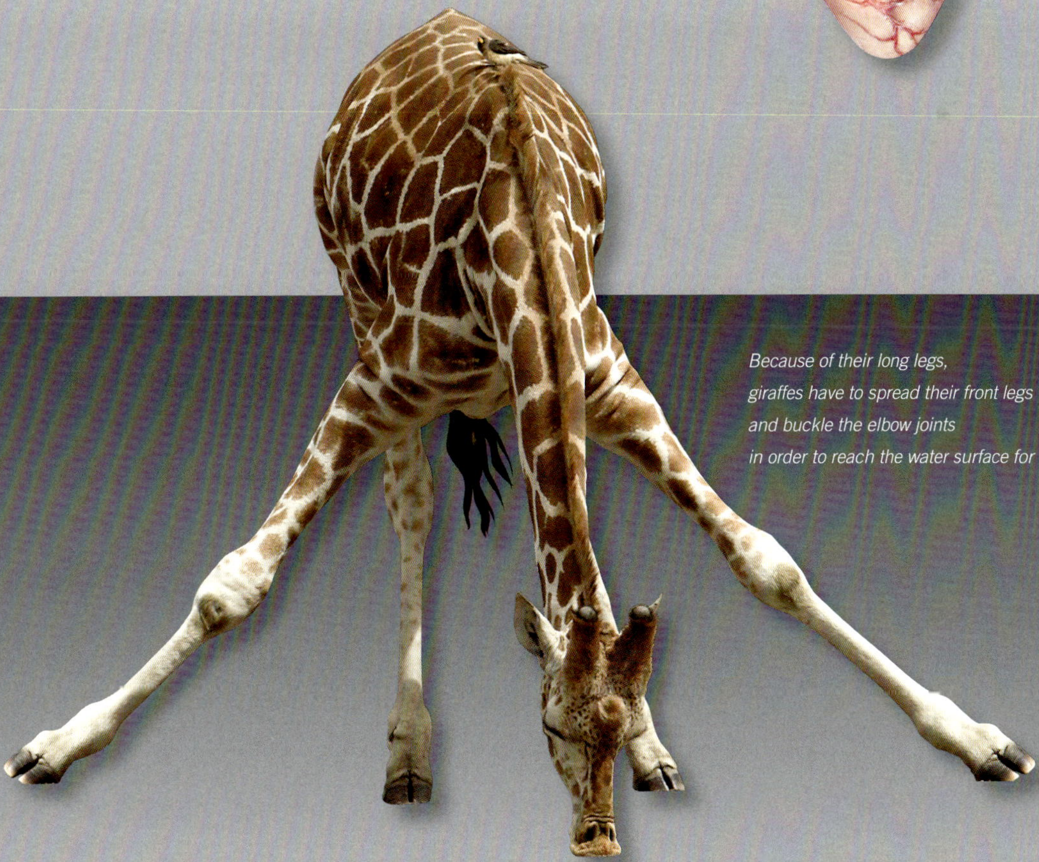

Because of their long legs,
giraffes have to spread their front legs far apart
and buckle the elbow joints
in order to reach the water surface for drinking.

Gentle
Giants

Elephant
(Elephas maximus indicus)

Elephants are the largest living land animals.
Asian elephants like this
measure up to 3.5 metres (11.5 ft) tall,
5.5 to 6.5 metres (18–21 ft) in length,
and weigh up to 5 tonnes.
Their ears are much smaller than those of their African cousins.

The most conspicuous anatomic feature of the elephant is its trunk.
It has evolved from the upper lip and the nose.
It contains neither a nasal bone nor any other bones, but consists solely of muscle tissue.
Thanks to about 40,000 muscles organised in bundles and a very sensitive innervation, the trunk is a versatile tool.
It serves as an organ of smell and taste, as a weapon, as a grasping hand while feeding, and as a suction and pressure pump while drinking.
With every gulp, the elephant can take up 8 to 10 litres (2–2.6 gal) of water into its trunk.

The internal organs of an elephant are, relatively speaking, no larger than in other mammals.
Its brain weighs about 4 to 5 kilograms (9–11 lbs).

The heart, depending on its age, weighs between 12 and 21 kilograms (26–46 lbs).
It beats about 30 times per minute.

Depending on their species, elephants have a skeleton of 326 to 351 bones that are moved by roughly 394 skeletal muscles.

The bones are extremely heavy and do not contain any marrow.
The legs are built very straight to carry the extreme weight.

Elephants usually amble, covering about 5 km/h (3 mph).
When in danger, they can run at speeds up to 40 km/h (25 mph).

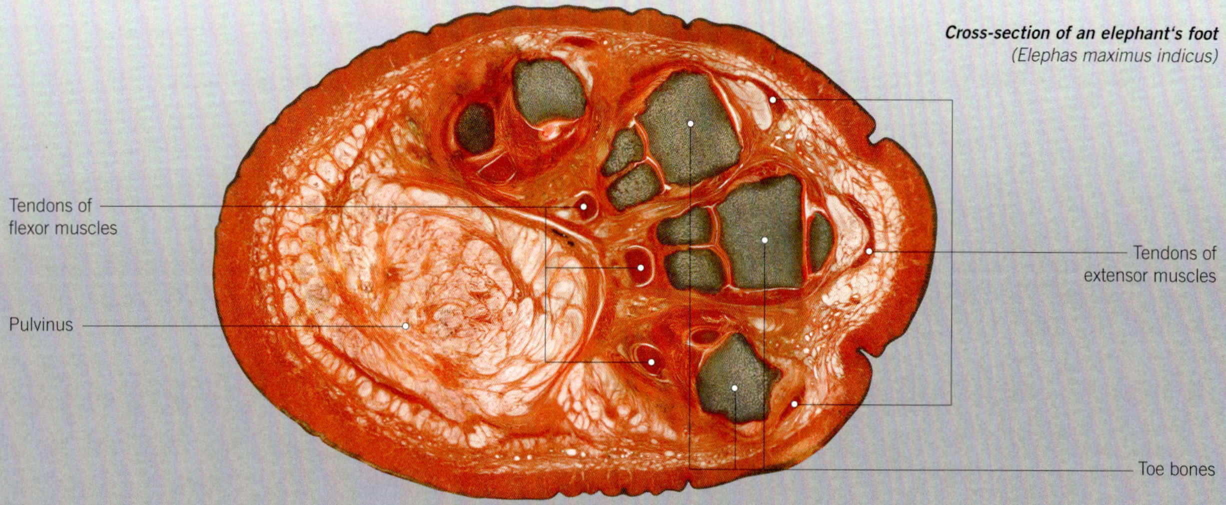

Cross-section of an elephant's foot
(Elephas maximus indicus)

Elephants seem to be plantigrades.
In reality, however, they are unguligrades, as can be seen in this longitudinal section.
The toe and foot bones are raised sharply, so that only the tips of the toes touch the ground.
The other bones rest on elastic, cushioned soles.
Because of the thick padding, the foot appears round and massive.
The ankle joints have a limited range of movement, which is one of the reasons why elephants cannot jump.

Digitigrades like cats and dogs touch the ground with all three phalanges, which make them quick and silent runners.
Plantigrades, such as bears and humans, walk on their entire foot, that is, with toes, metatarsals, and heels.

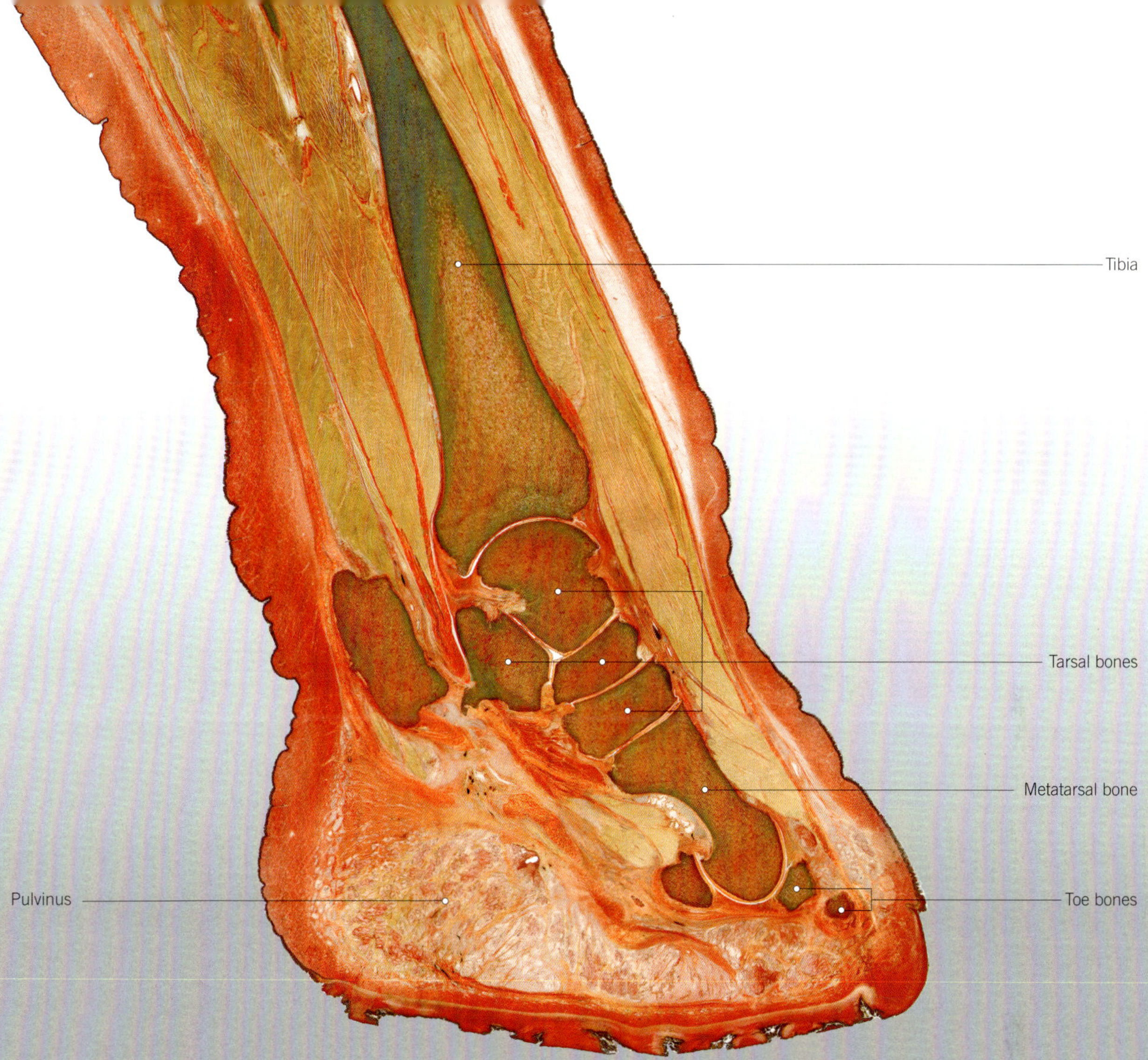

Longitudinal section of an elephant's foot
(Elephas maximus indicus)

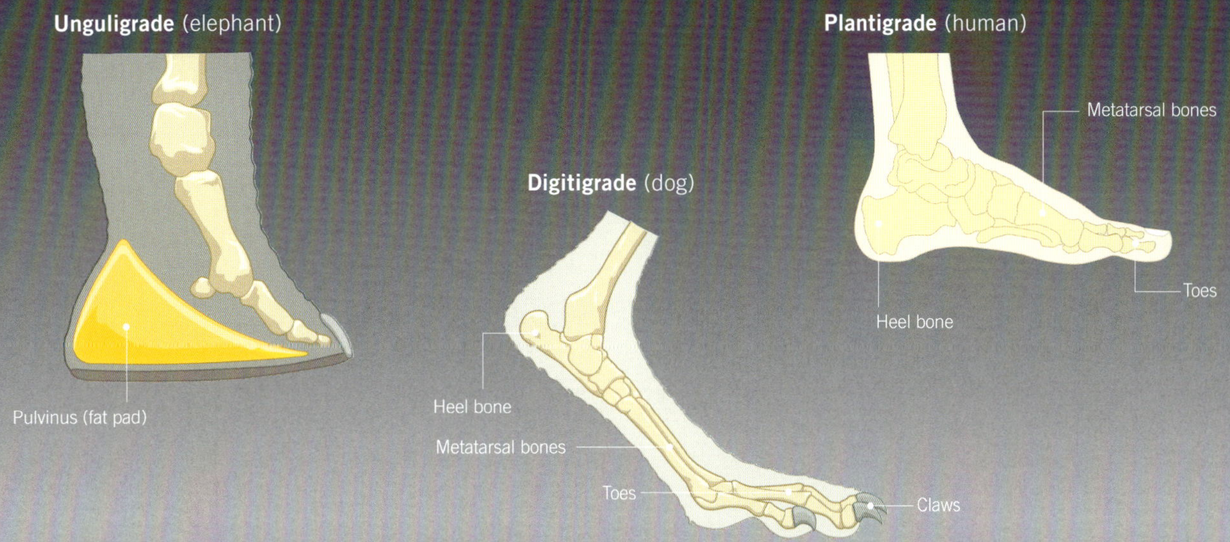

Elephants are pure herbivores.
They utilise only about 40% of their food intake, as they have a less efficient digestive system than ruminants.

Dissected elephant
Individual sections of its body have been expanded to offer a view of the spatial relationships between bones, muscles and organs.
(Elephas maximus indicus)

This body slice shows a
2-mm-thick (0.08 in) cross-section through the body of a female Asian elephant from trunk to tail.

Notable are the air-filled cavities in the skull bones that reduce the weight of the head. Nevertheless, the impressive head is of considerable weight and can account for up to a quarter of the total body weight. The bony struts give the skull bone the necessary stability to carry the trunk and tusks.

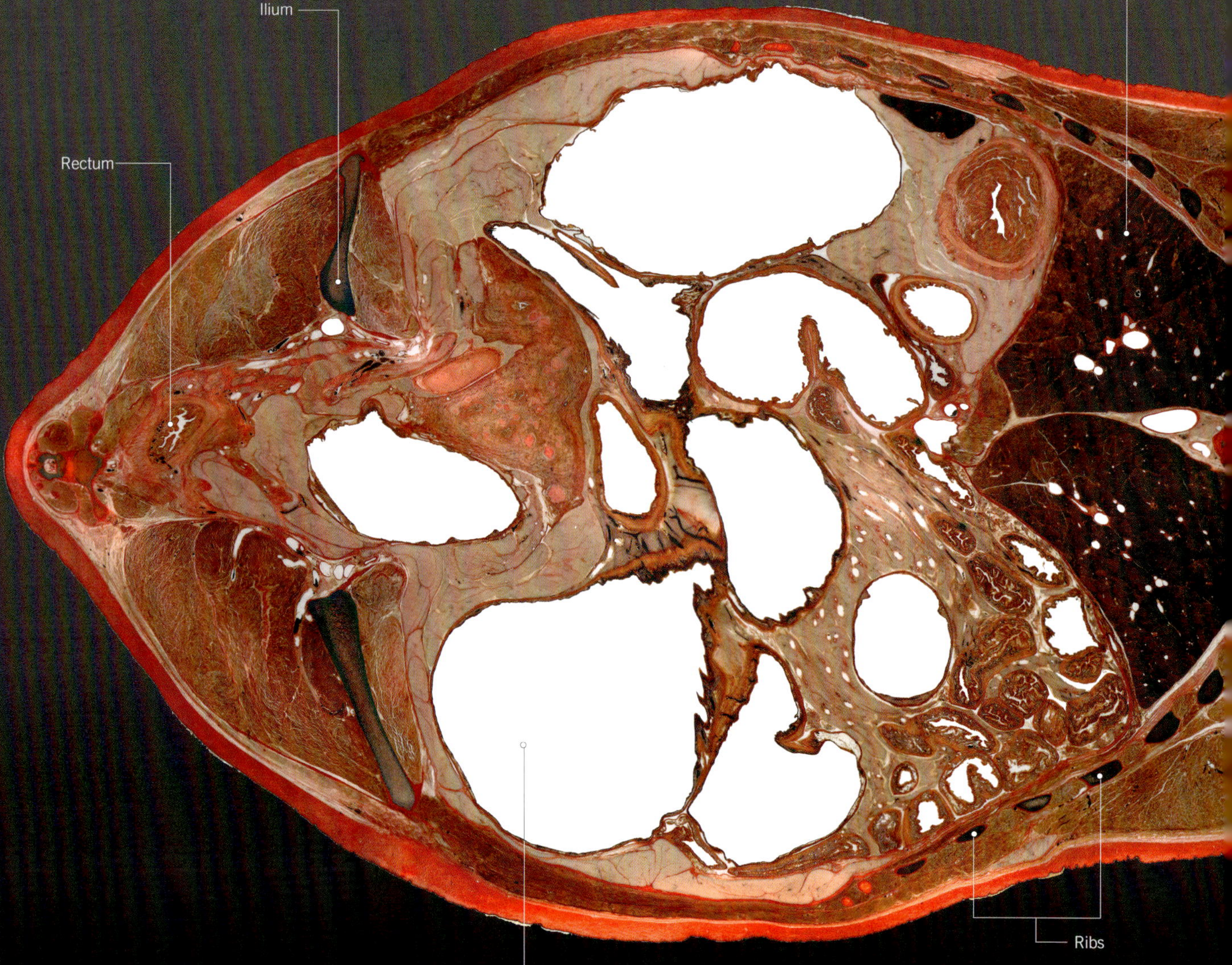

The two lungs can be seen in the chest cavity. Contrary to all other mammals, in elephants they are connected to the chest with connective tissue. This allows the elephant to cross a river and use its long trunk as a snorkel without suffering from the pressure difference between the atmospheric air and the lungs, despite being up to 2 metres (6.5 ft) under water.

The elephant's colon is very large like in any other herbivore. It fills almost the entire abdominal cavity and digests up to 200 kilograms (441 lbs) of plant food that elephants consume every day.

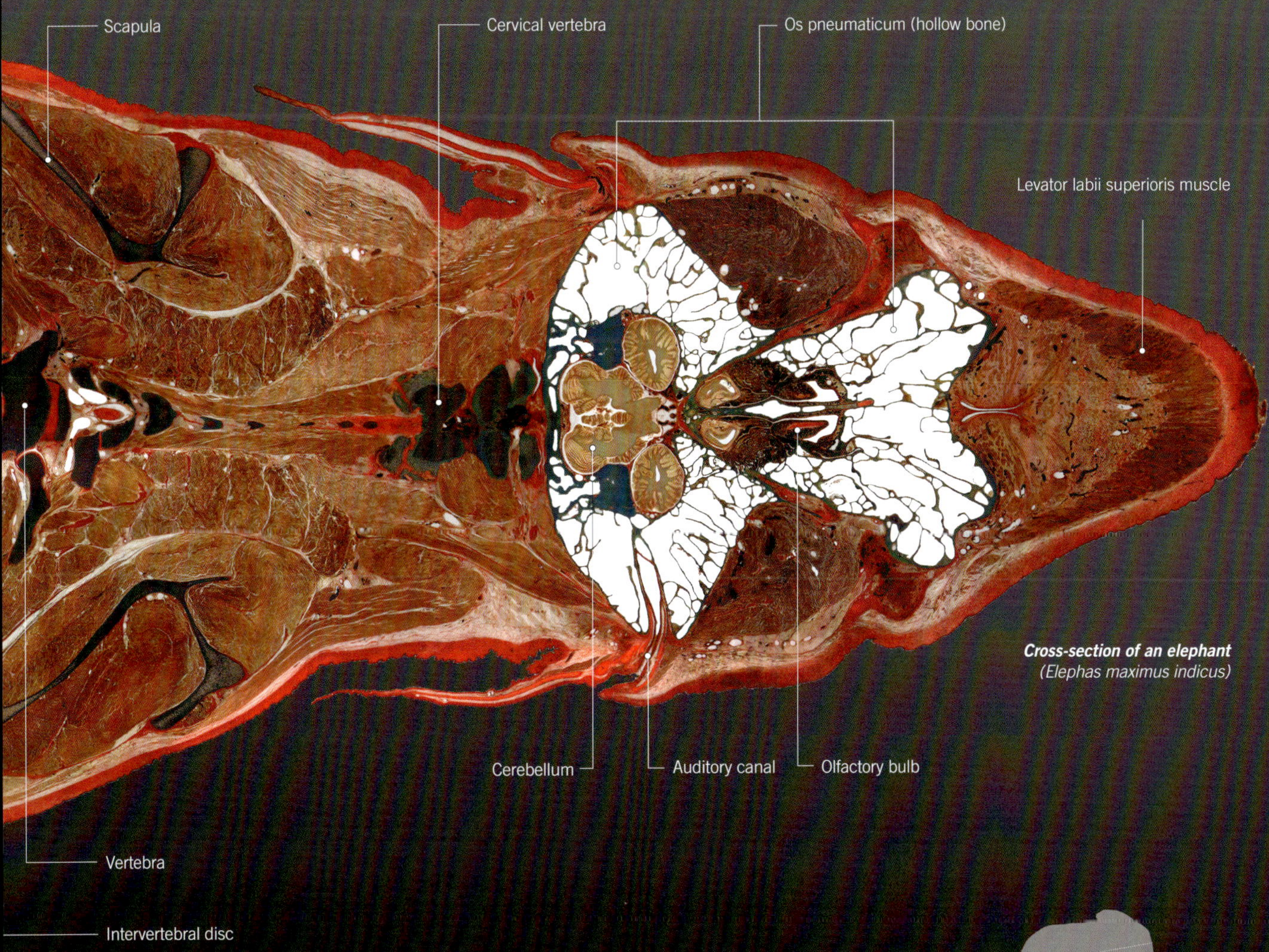

Cross-section of an elephant
(Elephas maximus indicus)

Almost Human

Lowland Gorilla
(Gorilla gorilla gorilla)

Other primates are the mammals
that most resemble humans in their anatomy
and behaviour.

This lowland gorilla weighs 200 kilograms (420 lbs)
and stands about 1.85 metres (6.1 ft) tall.
The circumference of its chest is 1.55 metres (5.1 ft),
and its stretched out arms span 2.4 metres (7.8 ft).
Female gorillas are shorter (1.3 metres / 4.3 ft)
and only weigh half as much.

Gorillas generally walk on all fours.
Their spine's curvature differs from that of humans
and makes standing upright strenuous for them.

The bones of their forearms, *metacarpi*,
and fingers are relatively long.
Thus, when standing,
the gorilla's arms extend down beyond their knees.
The configuration of arm and leg muscles in the gorilla
is very similar to that in humans.
One major difference is in the anatomy of the foot.
Unlike humans,
gorillas can also use their toes for grasping.

Lowland Gorilla
(Gorilla gorilla gorilla)

The Discerning One

Within the biological classification system, humans are a mammal in the order of primates. Genetic analysis has shown that bonobos (pygmy chimpanzees), chimpanzees, gorillas and orangutans are the closest relatives of humans.

Unlike all other primates, humans are capable of permanently maintaining an upright posture, because only humans are equipped with a double S-shaped curved spine. When walking upright, it acts like a spring and absorbs all the shocks and vibrations. In contrast, the entire body weight of great apes is supported by a simple curved spine. Due to this fact, they have to shift part of their weight to their arms and use them as support while walking.

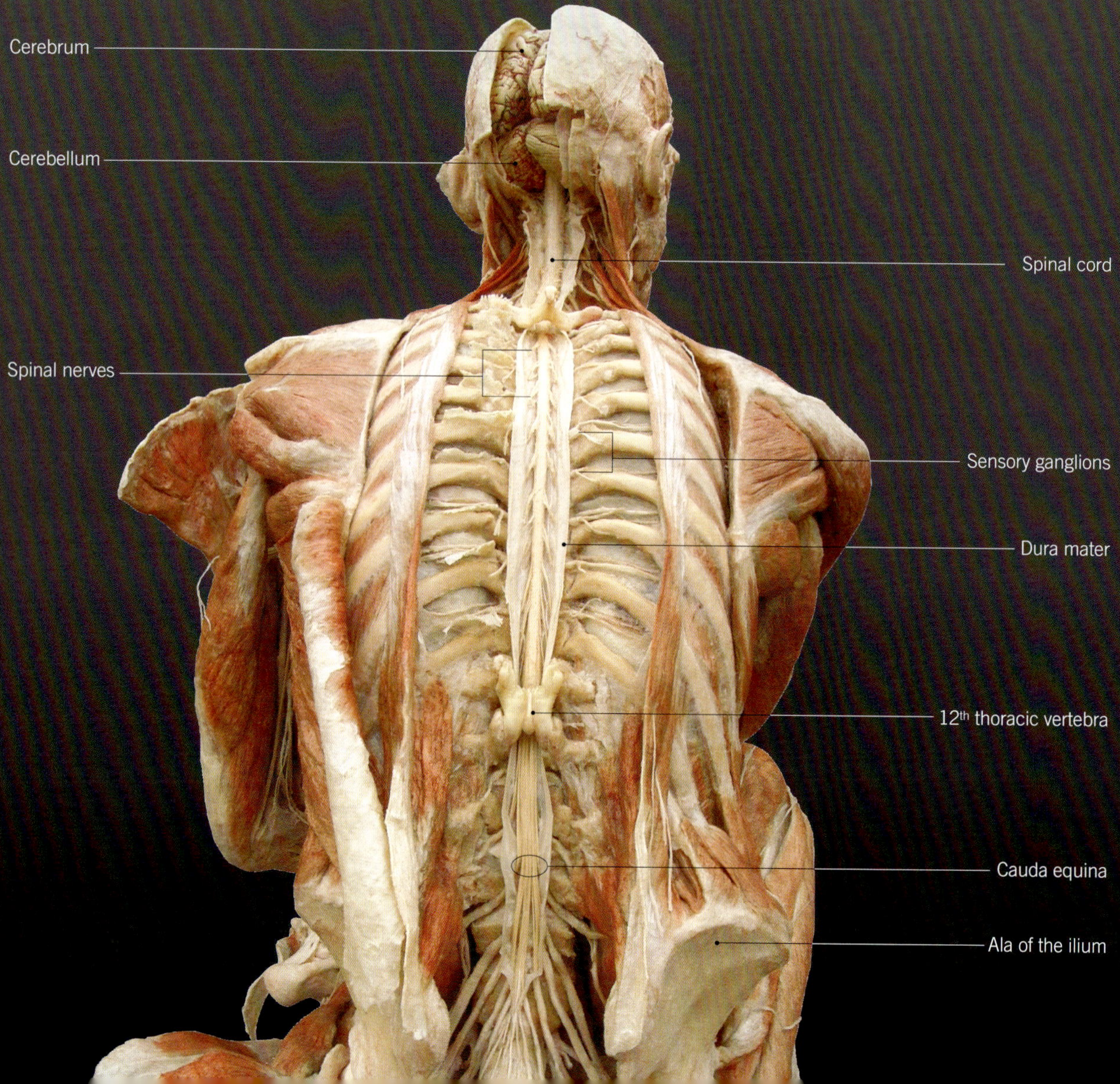

Particularly remarkable is the difference in size and brain performance. Humans have the capacity for self-awareness and reflection. They are also able to express and share complicated thoughts with their language and writing, remember past events, and plan for the future – skills that far exceed the capacities of apes.

Also, thanks to science and technology, humans have become increasingly independent from their environment.

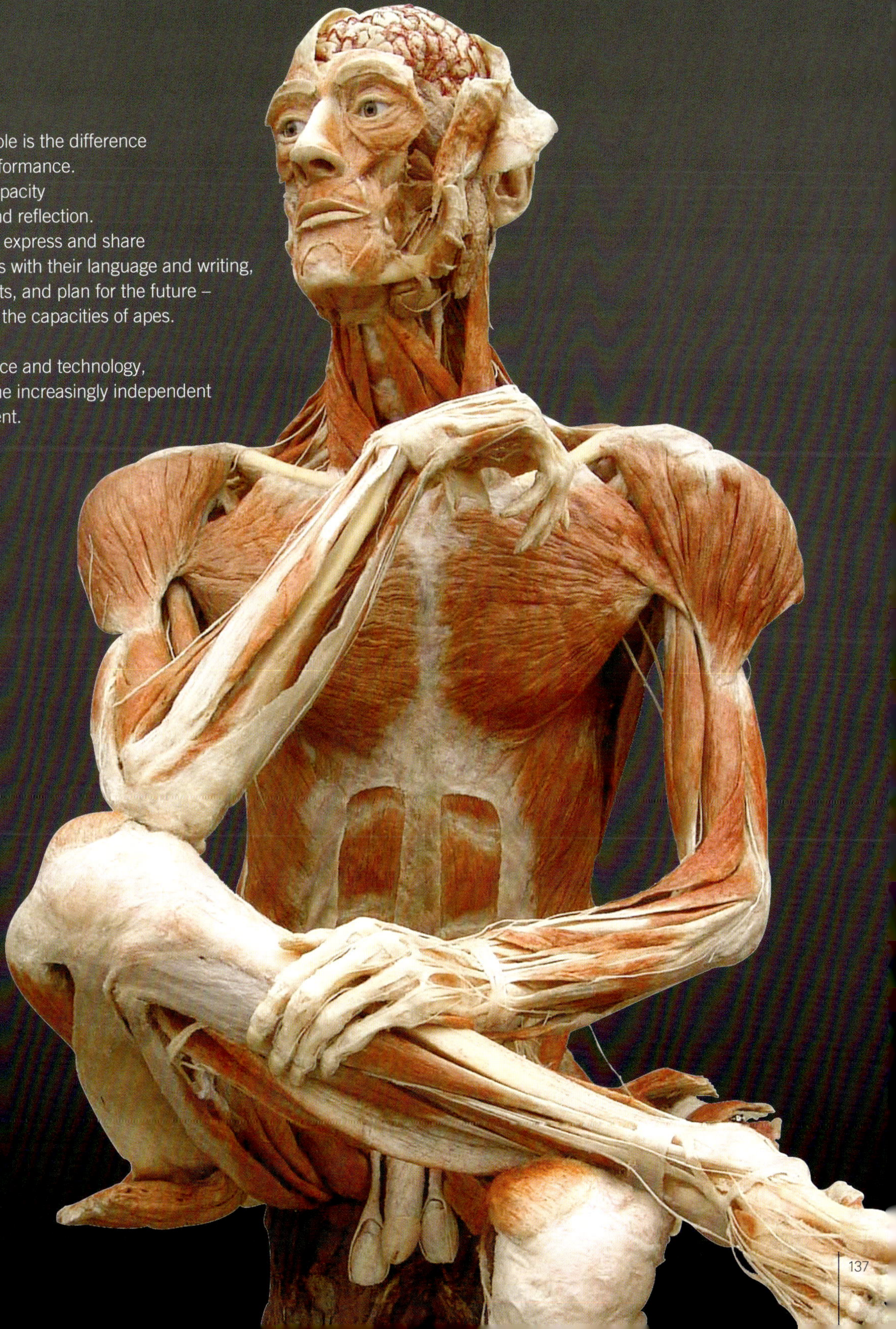

The Ponderer
(Homo sapiens sapiens)

Humans are anatomically similar
to most other mammals in regards to the form,
location and microscopic structure
of their organs and muscles.
However, large differences are often found
in the proportions.
For example, the muscular system of humans
is rather poorly developed compared to that of a horse.
On the other hand, the more developed brain of humans
makes them significantly more intelligent.

Major differences are also found in the extremities.
In humans, the upper arm and thigh bones
are relatively long,
while the hands and feet are rather short.
In horses the opposite is the case.
The upper arm and thigh bones are very short,
so that the elbows and knees are close to the torso.
The other arm and leg bones are relatively long,
especially those of the hands and feet.
Their length makes up about a third of each limb.
Also, the number of fingers and toes
is reduced to one digit.

Rearing Horse with Rider
(Equus ferus caballus, Homo sapiens sapiens)

PUBLISHING INFORMATION

CATALOGUE
ANIMAL INSIDE OUT, a Body Worlds Production
Angelina Whalley
Gunther von Hagens

EDITING
Angelina Whalley

DESIGN
DIE-**WERBE**AKTIVISTEN.de, mArc Schumacher, Mannheim, Germany

2st printing, revised edition

Arts & Sciences
Verlagsgesellschaft mbH, Heidelberg, Germany

ISBN 978-3-937256-31-3

www.AnimalInsideOut.com